W0082739

NOBEL LAUREATES of LOS ALAMOS

NOBEL LAUREATES OF LOS ALAMOS
THE MANHATTAN PROJECT ERA

Published for Los Alamos National Laboratory by

Texas A&M University Press
College Station

Copyright © 2023 by Texas A&M University Press
All rights reserved
First edition

This paper meets the requirements of ANSI/NISO Z39.48-1992
(Permanence of Paper).
Binding materials have been chosen for durability.
Manufactured in China by Martin Book Management.

Library of Congress Cataloging-in-Publication Data
Library of Congress Control Number: 2022948989
Identifiers: LCCN | ISBN 978-1-64843-163-0 (hardcover) |
   ISBN 978-1-64843-164-7 (ebook)
LC record available at https://lccn.loc.gov/2022948989

Los Alamos National Laboratory, an affirmative action/equal opportunity
employer, is operated by Triad National Security, LLC for the National
Nuclear Security Administration of U.S. Department of Energy under contract
89233218CNA000001. The U.S. Government retains nonexclusive, royalty-
free license to publish or reproduce the published form of this contribution,
or to allow others to do so, for U.S. Government purposes. This work was
performed under the auspices of the U.S. Department of Energy. Los Alamos
National Laboratory strongly supports academic freedom and a researcher's
right to publish; as an institution, however, the Laboratory does not endorse
the viewpoint of a publication or guarantee its technical correctness.

Atomic Weapons Establishment Historian Richard Moore's contributions
are U.K. Crown Owned Copyright and appear with the permission of the
U.K. Ministry of Defense.

LA-UR-22-24473

Contents

Foreword

There are 18 Nobel Prize-winning scientists

who worked at the secret wartime laboratory

of the 1940s. I knew three of them.

The scientists whose lives are chronicled in this book — 18 Nobel prize winners plus their one-time leader, J. Robert Oppenheimer — had crucial roles in the dawn of the Atomic Age. They were all part of the endeavor that launched it: the Manhattan Project. And beyond their contributions to the development of the atomic bombs that helped end World War II, they were pillars in the development of science and technology throughout the twentieth century. Each was awarded the Nobel Prize because of their compelling work. Their lives and accomplishments, which fill these pages, form part of the intellectual patrimony of humanity and this book preserves their memory.

I can say that with certainty, as a Los Alamos physicist who followed in their footsteps — first as a technical staff member in 1983, later as a group leader from 1994 to 2005, and then as the head of the Theoretical Division in 2006. Not long after I stepped down in 2015, I was honored to be appointed in 2018 as the division leader of the lab's Richard P. Feynman Center for Innovation. Before my retirement, I concluded my active career as a science advisor to

Antonio Redondo, physicist and retired head of the lab's Theoretical Division is pictured here in front of Nobel laureate Niels Bohr's tomb in Copenhagen, Denmark. (Photo courtesy of Antonio Redondo.)

Roy Glauber came to Los Alamos at the age of 18 to help develop the atomic bomb. (Photo courtesy of Jane Reed/Harvard University.)

Hans Bethe was the first physicist to lead the lab's Theoretical Division.

After World War II ended, Richard Feynman left Los Alamos to teach at Caltech, among other pursuits. (Photo courtesy of the U.S. Department of Energy/Wikimedia Commons.)

New Mexico's governor in 2020-2021. It's not an overstatement to say that with every job, I was standing on the shoulders of giants.

Furthermore, I was fortunate to have interacted with three of these giants. When I was a graduate student in physics at Caltech, I took a year-long course in advanced quantum mechanics from Richard Feynman (see Page 70) in 1972. He was a wonderful teacher, always full of enthusiasm that permeated all the students, even those who weren't studying science. I remember arriving on Day 1 to a crowded classroom, overflowing beyond capacity with what were obviously students who would not have been eligible to enroll in his course. They were there just to hear what Feynman had to say.

Years later, I became the division leader of the Theoretical Division and was honored to have the same position first held by Hans Bethe (see Page 22). I had met him during a couple of his visits to the laboratory before his death in 2005, and though he was getting on in years, he was just as lively and sharp as any young scientist there. It was a lot of fun to talk to him. He was quite interested in the things we were doing in the division and asked challenging questions.

My connection with Roy Glauber (see Page 84) was a fortuitous meeting at the Lindau Nobel Laureate Meeting in 2016. I was a panelist and Glauber was one of the speakers. A friend introduced us and Glauber reminisced about coming to Los Alamos as an 18 year old during the Manhattan Project years. He had many anecdotes and insights, some of which I was not aware of until reading this book.

These men are the focus of just three among more than a dozen fascinating profiles that weave personal histories, tragedies, triumphs, science, and a lab in northern New Mexico into one remarkable book. I invite you to immerse yourself in their stories and am certain you will be as inspired by these people as I always have been.

Antonio Redondo
Los Alamos

Acknowledgments

Nobel Laureates of Los Alamos:

The Manhattan Project Era is a classic

example of a production by diverse hands,

an archaic phrase that means a project

created by many talented individuals who

come together as one.

Though many people had a hand in the creation of this book the two who had the greatest impact were Brye Steeves and Paul Ziomek. Brye served as the Managing Editor of the project, coordinating with every person on this book, as well as editing all the content and writing some of it, too. Paul conceptualized and implemented the unique aesthetics of the book, overseeing quite literally every visual aspect that fills these pages.

I drew the authors of this book from staff in the National Security Research Center (NSRC). I had no shortage of volunteers eager to contribute. This book's writers ranged from our historians to our archivists to our librarians to a trio of our professional editors.

The following individuals contributed the biographies and other sections in this book: Amy Belotti, Alan Carr, Hadley Hershey, Megan Jochem, Jackie Kilby, Nic Lewis, Mott Linn, Ellen McGehee, Laura McGuiness, Roger Meade, Renae Mitchell, John Moore, Richard Moore (of the U.K.'s Atomic Weapons Establishment), Octavio Ramos, Brye Steeves, and Madeline Whitacre.

The three editors that worked cover-to-cover on this book were Renae Mitchell, Octavio Ramos, and Brye Steeves.

The distinctive styling and artwork of this book was custom-created by the NSRC's design team, led by Paul Ziomek. Gabriella Smith assisted Paul, and original illustrations were provided by Mike Nudelman.

In addition to the authors whose names appear on these stories, there was a sizable cadre of people behind the scenes who helped develop each one. Every group within the NSRC contributed to this book: the publishing team, the archivist team, the digitizing team, the historian team, and the librarian team. These highly skilled, specialized staff helped us find the one-of-a-kind documents and photos contained in this book, as well as mined for the details that bring each character to life.

The following individuals served as content consultants for this book: Alan Carr, Alan Hurd, Nic Lewis, Ellen McGehee, Roger Meade, and Tony Redondo. Collections support was provided by Danny Alcazar, Patricia Cote, John Moore, and Angie Piccolo.

The following staff reviewed the entire book for security classification and cleared it for publication: Danny Alcazar, Chris C'de Baca, and Andrew Gordon.

Rizwan Ali
Editor-in-Chief

NOBEL LAUREATES of LOS ALAMOS

Introduction

Los Alamos National Laboratory has always attracted the most brilliant people from around the world. This legacy started at the inception of the lab during World War II, as part of the Manhattan Project, and continues to this day.

For those of you wanting some proof for this claim, I present you this book, which provides a glimpse into the lives of 18 impressive minds that worked at Los Alamos during the Manhattan Project.

This book stemmed from a series of conversations I had with the lab's Senior Historian, Alan Carr, who came up with the idea of a collection that encapsulates the scientists who worked in Los Alamos as part of the Manhattan Project and either won or would subsequently win the Nobel Prize. The book you hold in your hands is the result of this idea and our conversations, one that chronicles the individuals who won the prize in physics, chemistry, and peace.

It Began with Nobel himself

Like many of the personalities found in this book, Alfred Nobel was a renaissance man, a multifaceted individual who one moment was busy inventing something revolutionary and the next developing a new business model or writing a poem or an act in a play. A holder of 355 patents, Nobel is most remembered as the inventor of dynamite.

Nobel's varied interests led him to create the Nobel Prize, which he never saw come to fruition. Upon his death, he bequeathed much of his wealth ($3.1 million, valued at $265 million today) toward establishing the prize, which would honor individuals for exemplary achievements in physics, chemistry, physiology or medicine, literature, peace, and, later, economics. The first Nobel Prize was awarded in 1901, four years after Nobel's death.

When initially visualizing this book, I wanted the authors of these biographies to paint fully realized portraits of these 17 men and one woman, bringing them to life for readers. Thus, this book is more than just a collection of the scientific advancements made by these incredible people. Each biography here covers a wider canvas, one that attempts to capture the personality behind these incredible scientists.

Herein you will find a practical joker, a stubborn but wise leader, an avid mountaineer and fly-fisherman, an introspective beach walker, a gifted baritone, an everyman humorist, and a flamboyant communicator. The three-dimensionality of these individuals, warts and all, makes them more than Nobel laureates — it makes them human.

Worth mentioning is that the laboratory's first director and famed physicist J. Robert Oppenheimer never won a Nobel Prize, despite his groundbreaking work on nuclear fission and the conceptualization of black holes, among other notable achievements (see Page 96 and read why Oppenheimer never won). Regardless, he too was an exceptional person whose legacy certainly endures.

In Closing

Every morning that I come to work at the National Security Research Center, the laboratory's classified library, I am amazed by the millions of documents, films, books, and other materials related to science, engineering, and history that span the entire breadth of nuclear weapons. These national treasures served as valuable points of research for this book's writers and yielded the many photographs and documents that fill its pages. Unless otherwise noted in the captions, these visual elements are from the collections of the National Security Research Center.

It is fitting that the National Security Research Center developed this book, since our lineage goes back to the technical library formed by J. Robert Oppenheimer in 1943 at the beginning of the Manhattan Project. Ours is the first to chronicle Nobel laureates with a World War II connection to what is today known as Los Alamos National Laboratory. I hope you enjoy this book as much as I did overseeing its development.

Rizwan Ali
Editor-in-Chief
Director, National Security Research Center
Los Alamos National Laboratory

A retired colonel from the United States Air Force, Rizwan "Riz" Ali has accumulated over three decades of experience in nuclear operations, cyber security, information technology, and space systems. Ali is also an engineer, a published writer, and an avid photographer. He helped establish the National Security Research Center and serves as its first director.

THE NOBEL PRIZE AT A GLANCE

BY BRYE STEEVES

WHAT IS THE NOBEL PRIZE?

The Nobel Prize is globally accepted as the **PINNACLE** of recognized achievement, and is one of the most prestigious honors an **INDIVIDUAL** or **ORGANIZATION** can receive.

The Prize is awarded in six areas: **PHYSICS, CHEMISTRY, MEDICINE, LITERATURE, ECONOMICS,** and **PEACE.**

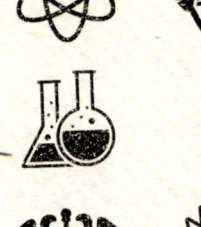

Recipients receive a **GOLD MEDAL** (see Page 13); a prize document; and a monetary award, which, in 2021, was about **$1.14 MILLION.**

4

WHO ESTABLISHED THE PRIZE?

The prize was conceived by scientist and entrepreneur **ALFRED NOBEL** (see Page 34). Nobel established the prize in his will, though it was **FIRST** awarded in **1901**, five years after his death.

Prizes are bestowed on "those who, during the preceding year, shall have conferred the greatest benefit on mankind."

A Nobel prize cannot be awarded to more than three individuals per category, but the Nobel Peace Prize can be awarded to organizations. Prizes are not awarded posthumously. As of 2021, **943** individuals and **25** organizations have received Nobel prizes.

HOW IS THE PRIZE AWARDED?

The Nobel committee considers **THOUSANDS** of **NOMINATIONS** before choosing about **300** potential **RECIPIENTS.**

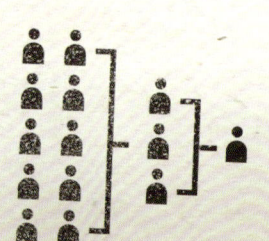

Next, a report with experts' commentary is sent with a list of preliminary candidates to prize-awarding institutions, which choose the laureate or laureates in each field based on a **MAJORITY** vote.

WHERE IS THE PRIZE AWARDED?

The Nobel Prizes are presented at ceremonies in **STOCKHOLM, SWEDEN** (Alfred Nobel's birthplace), with the exception of the Nobel Peace Prize, which is presented in **OSLO, NORWAY,** though no one knows why.

Oslo, Norway

Stockholm, Sweden

WHEN IS THE PRIZE AWARDED?

The Nobel Prizes have been awarded **ANNUALLY** since 1901,

with the exception of the prize in **ECONOMICS,** which was added by the **BANK OF SWEDEN IN 1968.**

WHY WAS THE PRIZE ESTABLISHED?

Alfred Nobel ENVISIONED the NOBEL PRIZE as a means to posthumously use his vast fortune to honor significant ACHIEVEMENTS on a GLOBAL SCALE.

The Nobel Prize medal, given to each recipient, features an image of Alfred Nobel, its benefactor and namesake. Frederick Reines (see Page 124) was awarded the Nobel Prize in Physics in 1995 for the detection of the neutrino. Laureates can order up to three replicas of their Nobel medal; one of Reines' replica medals, pictured here, is part of the lab's National Security Research Center's historic collections.

LUIS ALVAREZ

By Laura McGuiness

Luis Alvarez can be thought of as the King Midas of physics.

But instead of turning objects into gold like the character fabled in Greek mythology, Alvarez turned some of science's biggest questions into physics' greatest discoveries. This pattern is exemplified in the countless endeavors that he undertook.

In the 1950s, he helped the Central Intelligence Agency (CIA) determine that UFOs were not a potential security threat. A decade later, he investigated the Kennedy assassination, mathematically proving that the sudden movement of the president's head indicated a bullet was shot from behind him. Alvarez was even involved with searching Egyptian pyramids for unknown chambers and, later in life, postulated that the extinction of dinosaurs was caused by the impact of a large asteroid. And at the once-secret Los Alamos lab, Alvarez developed detonators for the plutonium implosion weapon known as Fat Man that would help end World War II.

Beginnings

Born in San Francisco, California, on June 13, 1911, Alvarez's own genius must have been practically predictable based on his gifted lineage. He was the son of a successful U.S. physician and the grandson of a Spanish physician responsible for diagnosing leprosy.

As a child, Alvarez lived wherever his father went for work. They moved from California to Cananea, Mexico, and finally to Rochester, Minnesota. Alvarez attended the University of Chicago, where he rapidly completed an education in physics, earning a bachelor's degree in 1932, a master's degree in 1934, and a Ph.D. in 1936.

Luis Alvarez (center) poses with Lawrence Johnston (left) and Harold Agnew (right). All three men flew on the B-29 observation aircraft called The Great Artiste as scientific observers responsible for ensuring the proper use of the blast gauge canisters during the detonation of Little Boy.

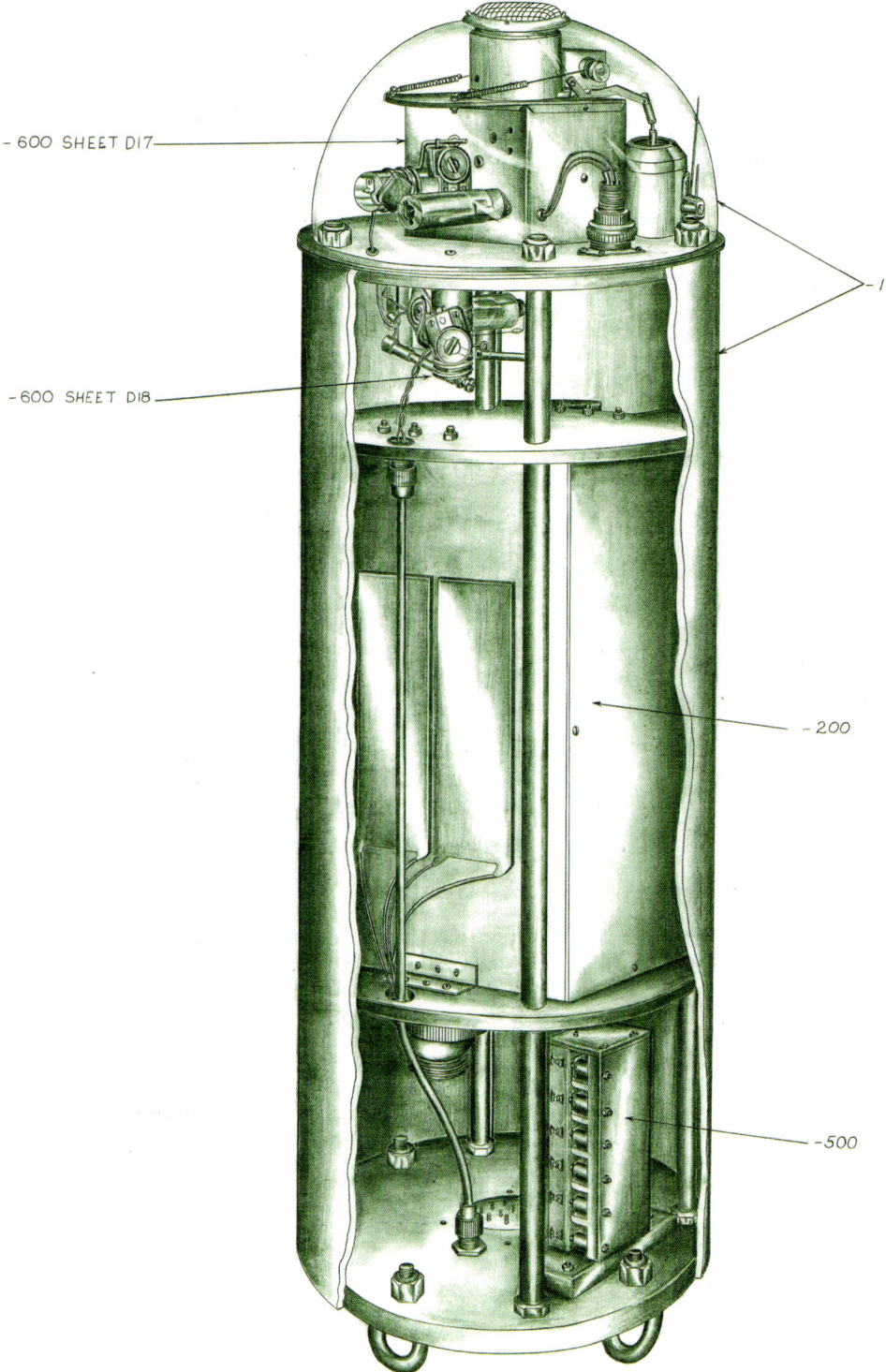

- 600 SHEET D17

- 600 SHEET D18

- 100

- 200

- 500

Y-2450-DI

AIRBORNE CONDENSER GUAGE
ASSEMBLY OF GUAGE

F.D.BEMIS – 9-3-46

Early career and a Los Alamos offer

Following his education, Alvarez married and then moved to Berkeley, California, where he worked under nuclear scientist and 1939 Nobel laureate Ernest Lawrence at the University of California, Berkeley Radiation Laboratory. Throughout the pre-World War II years, Alvarez contributed to a number of radar projects both in the United States and abroad. The radar-related idea for which Alvarez is best known is ground-controlled approach (GCA). While watching a ground-based, anti-aircraft fire-control radar track an airplane, it occurred to him that "if a radar could continuously and automatically track an enemy aircraft accurately enough to shoot it down, the same information should be adequate to guide a friendly pilot to a safe landing in bad weather," according to his autobiography. GCA is still used today and the concept served as the basis for Arthur C. Clarke's novel, *Glide Path*, which includes a character modeled after Alvarez.

By the fall of 1943, Alvarez received an offer from famed physicist J. Robert Oppenheimer to work in Los Alamos as part of the Manhattan Project — the U.S. government's war time effort to create the world's first nuclear weapons. Alvarez arrived on April 25, 1944 and began working under physical chemist George Kistiakowsky in the Applied Physics Division, a new explosives team created to perfect the plutonium weapon.

A diagram of a blast gauge canister, which carried the transmitters that Luis Alvarez developed for measuring blast-wave strength. Concurrent with the release of the Little Boy and Fat Man atomic bombs, three canisters were also deployed. As they floated under their parachutes, information about the blast was transmitted to scientific observers, including Alvarez.

Generating a spherical implosion around a plutonium core had eluded most Los Alamos scientists. According to Alvarez's memoir, he was armed with a hunch and a trusty graduate student assistant when he thought to use a large capacitor to detonate explosive charges within microseconds of each other. These nearly simultaneous detonations produced the pressure waves required to compress the plutonium core to the required density, leading to the success of the implosion-type nuclear weapon known as Fat Man.

Entering the Atomic Age

Before the weapon's test in the New Mexico desert in July of 1945, Alvarez asked Oppenheimer if there were any assignments that would take him to the Pacific. Oppenheimer knew just the one and Alvarez was tasked with developing a method that would measure the energy released by Little Boy and Fat Man upon detonation. Typically, a new weapon would be tested before use in combat, allowing scientists to collect all pertinent blast data. There was only one uranium weapon, however, and production for another plutonium weapon would not be possible under the lab's fast timeline. Instead, Alvarez created a different method of blast wave measurement that could be used during active combat.

A set of calibrated transmitters were outfitted with a parachute harness and, in tandem with the dropping of Little Boy and Fat Man, the transmitters would be parachuted out of the aircraft. As the aircraft flew away from the area of release, their instruments would receive the signal from the transmitters, allowing them to measure and record the strength of a blast wave from a safe distance. Before they could be used during active combat, however, they first had to be tested.

On July 15, 1945, the morning before the Trinity test (the first-ever detonation of an atomic bomb), Alvarez left his home in Los Alamos for the Alamogordo Bombing Range about 210 miles away. Alvarez wrote in his autobiography that he told his wife Geraldine "what most Los Alamos wives were told, that if she happened to be up at five o'clock the next morning and looked south, she might see something interesting." Tasked with ensuring the proper use of the transmitters, Alvarez flew as a scientific observer on an aircraft during the test.

Shortly after the Trinity test in the New Mexico desert, Alvarez and a small group of scientists were flown to Tinian island in the western Pacific Ocean in preparation for the detonation of Little Boy. For safety reasons, Oppenheimer had initially refused to allow scientific observers to fly into combat, but Alvarez campaigned for himself and two other scientists to fly with the equipment. Oppenheimer eventually relented. Once again tasked with ensuring the proper use of the transmitters, Alvarez flew as a scientific observer during the release of Little Boy above Hiroshima, Japan, on August 6, 1945.

Japan surrendered and World War II officially ended on September 2, 1945. Alvarez and his wife and two children moved back to Berkeley, where he had accepted a professorship. During this period of his life, he divorced his wife of 21 years, remarried several years later to Janet Landis (also an employee at the Radiation Lab), and fathered two more children.

Alvarez was even involved with searching Egyptian pyramids for unknown chambers and, later in life, postulated a controversial theory for the extinction of dinosaurs.

Luis Alvarez (right), pictured on Tinian island, posing with Fat Man's plutonium core. The B-29 military aircraft that released Little Boy and Fat Man above Japan took off and landed at Tinian island.

CN86406

Luis Alvarez (upper right) and scientists Harold Agnew (upper left), Lawrence Johnston (bottom left), and Bernard Waldman (bottom right) developed the canisters that recorded blast data from Little Boy and Fat Man. The atomic bombs were released in August 1945.

Prize-winning work

During the post-war period, Berkeley's Ernest Lawrence asked Alvarez to promise that he would complete the development for a hydrogen bubble chamber. In his memoir, Alvarez seemed mildly taken aback by the implications of this promise. He admitted he did "flit from one project to the next," but rarely left any of them before success was assured. In asking for Alvarez's promise, Lawrence may have had the premonition that completion of the hydrogen bubble chamber would mean great things. It in fact won Alvarez the 1968 Nobel Prize in Physics.

The bubble chamber was originally invented in 1952 by Donald Glaser, who was subsequently awarded the 1960 Nobel Prize for his work. Glaser's bubble chamber is a means of visualizing particle tracks through the superheating of ether. After seeing the small (1-centimeter by 2-centimeter) cylinder in action, Alvarez realized he could improve it by using liquid hydrogen instead of ether. His team built a 7-foot-long metal chamber with large glass windows that allowed for increased visibility of the particle tracks.

As a particle passed through the liquid hydrogen and left in its wake a trail of bubbles, Alvarez and his team could photograph the path of the particle. These photographs were subsequently studied and measured, resulting in the discovery of a large number of previously unknown "fundamental particle resonances."

As stated in a speech about Alvarez and his work at the Nobel Prize ceremony in 1968: "Practically all the discoveries that have been made in this important field of high-energy physics have been possible only through the use of methods originated by Professor Alvarez."

Twenty years after accepting the Nobel Prize, Alvarez died on September 1, 1988 at the age of 77 from complications related to esophageal cancer. He was survived by his second wife, Jean; four children, Walter, Donald, Jean, and Helen; and countless contributions in the field of physics.

This letter was written by Luis Alvarez and addressed to Professor Ryōkichi Sagane, who was one of many nuclear physicists of international repute in Japan. The letter was dated August 9, 1945, and warns Sagane of the atomic bomb, urging him to implore Japanese leaders to surrender. Sagane did not receive the letter until October, after two atomic bombs were released and World War II had ended. (Luis W. Alvarez Letter to Ryōkichi Sagane; Manuscripts, Archives, and Special Collections; Washington State University Libraries.)

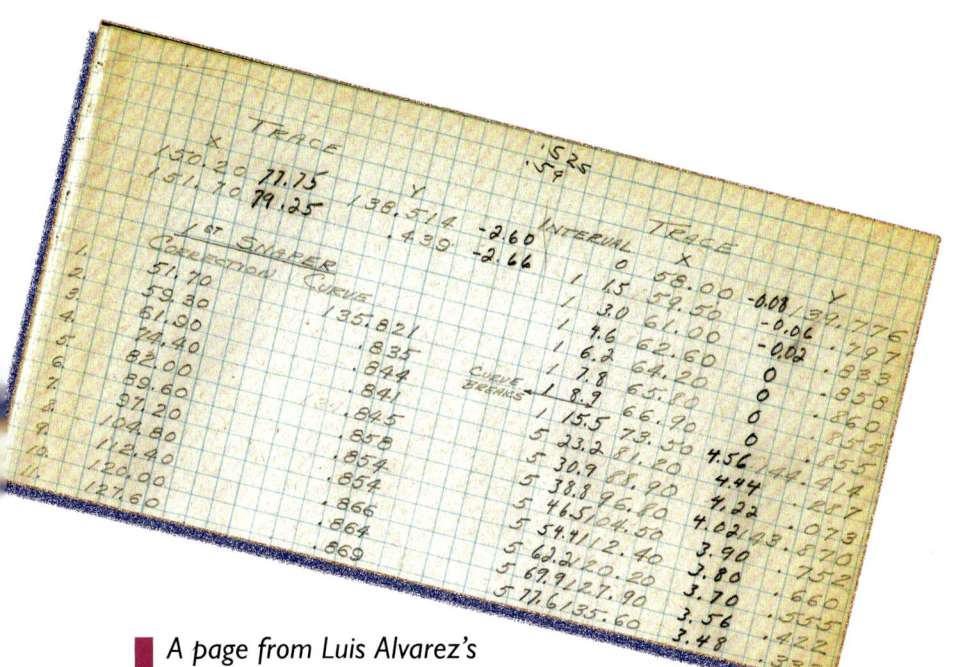

A page from Luis Alvarez's notebook shows his calculations of the first estimated yield produced by the Little Boy bomb (about 15 kilotons). On August 7, 1945, the day following the release of Little Boy over Hiroshima, Alvarez completed calculations based on the data received from one of the blast gauges that he and his team developed.

Luis Alvarez's Los Alamos lab badge photo was issued after his arrival on April 25, 1944.

LUIS ALVAREZ

1911–1988

LOS ALAMOS CONTRIBUTIONS

Worked on the lab's explosives team, where he suggested the use of a symmetrical implosion, leading to the success of the implosion-type Fat Man weapon. Also developed a set of calibrated transmitters to measure blast wave strength from an atomic explosion, which allowed the blast effects of Little Boy and Fat Man to be calculated.

NOBEL PRIZE: PHYSICS, 1968

Repurposed the bubble chamber, using liquid hydrogen and glass windows to capture photographs of particle resonance.

HANS BETHE

By Laura McGuiness
and Amy Belotti

How and why do stars shine? Hans Bethe earned a Nobel

Prize for answering that exact question — how stars live

and die. Fittingly, Bethe was a star himself, and not just because of the

prestigious prize he would later win.

A brilliant theoretical physicist, Bethe's work at Los Alamos contributed to the implosion method used in the two Fat Man-type bombs, first during the Trinity test in the New Mexico desert and in a second weapon that was released above Nagasaki, Japan.

A member of the Manhattan Project, which was the U.S. government's top-secret effort to create the world's first atomic weapons, Bethe served as the head of the Theoretical Division at the Los Alamos lab.

During this period, Bethe completed theoretical work that studied the hydrodynamics, that is, the motion of melted metals, to prove the feasibility of the implosion-type weapon codenamed Fat Man. The action of pushing the fissile material to achieve the critical mass (the minimum amount needed for a nuclear weapon to detonate) is called implosion. The forces and temperatures are very high, so the uranium or plutonium melts in the process, becoming liquid, hence the need to understand the hydrodynamics of these fluids.

He also developed the models that determined how neutrons diffuse through a critical mass, that is, how the neutrons move through materials (such as uranium or plutonium) to trigger an explosion. This, in turn, proved the feasibility of the gun-type weapon known as Little Boy. It was the first atomic bomb to be used in combat when it was released above Hiroshima, Japan, on August 6, 1945.

Later, Bethe's firsthand experience with weapons creation led to involvement with the President's Science Advisory Committee, which was created in 1957 by Dwight Eisenhower, where Bethe advocated for nuclear disarmament.

An early genius

Bethe was born on July 2, 1906 in Strasbourg, Germany. As a child he was interested in numbers and his mathematical abilities manifested themselves very early. By the age of 4, Bethe was pondering divisible numbers. At the age of 5, he was using chalk outside his house to calculate the square roots of numbers. By the age of 9 he was creating a table of base numbers and exponents.

Hans Bethe served as head of the laboratory's Theoretical Division, carrying out scientific research and addressing concerns about nuclear weapons.

As a young man, Bethe enrolled in the University of Frankfurt to study chemistry. He abandoned this field of study shortly after accidentally destroying his lab coat with sulfuric acid and deciding he was simply a poor experimentalist — a decision that pointed him firmly in the direction of theoretical physics, which relies on mathematics, among other disciplines. He entered the University of Munich in 1926, where he later completed his doctorate. Throughout the pre-World War II years, Bethe earned a host of fellowships at European universities.

Pre-war turmoil

In the early 1930s, however, anti-Semitism began to affect Bethe's work. He was born to a Jewish mother and a Protestant father, and although Bethe was raised under his father's religion and did not consider himself Jewish, his lineage would soon be problematic with Adolf Hitler's rise.

In 1932, Bethe was dismissed from his job as an assistant professor at the University of Tubingen in Germany due to an anti-Semitic law passed by the Nazi Party. Shortly thereafter, Bethe left the country, first emigrating

One of the most iconic buildings of the World War II period are Quonset huts, which were used as living quarters and work spaces for scientists to practice the assembly of the Fat Man weapon.

to England and then to the United States to pursue a career at Cornell University in New York. There, Bethe joined a group of other scientists working in the new physics department, including Stanley Livingston and Robert Bacher. Together, they published a series of nuclear physics articles — a compendium that was nicknamed "Bethe's Bible" and quickly became a standard resource for nuclear physics.

Prize-winning work

In 1938, Bethe initially declined a conference invitation because its topic of stellar energy generation did not immediately interest him. His last-minute decision to attend, however, went on to change both his own life and the field of physics.

At the conference, Bethe determined that a series of nuclear reactions were responsible for the sun's ability to shine day after day. Upon his return to Cornell, he continued to study these specific nuclear reactions, leading to the discovery of the carbon-nitrogen-oxygen cycle (CNO cycle).

The CNO cycle is significant because it demonstrates a process that converts hydrogen to helium. This process consists of a series of reactions which, together, produce enough stellar energy for the sun to burn for billions of years. The CNO cycle was considered a breakthrough in stellar nucleosynthesis and went on to win Bethe the Nobel Prize in Physics in

1967. Perhaps more importantly, however, a paper he wrote on the discovery was submitted to the New York Academy of Sciences and won a $500 prize — money that was used to help his Jewish mother flee Nazi Germany and immigrate to the United States.

On the mesa

In 1941, Bethe was invited by physicist J. Robert Oppenheimer to meetings on the initial designs of an atomic bomb. Following his discovery of the CNO cycle, Bethe was one of Oppenheimer's first recruits to the secret Los Alamos lab to create the world's first nuclear weapons.

Bethe initially resisted Oppenheimer's invitation, expressing both doubt about the atomic bomb's feasibility and concern for the moral implications of using such a weapon in combat. Ultimately, Bethe joined the project because he feared Nazi Germany was also trying to create a nuclear weapon.

As the head of the lab's Theoretical Division, Bethe worked to address concerns regarding the creation of nuclear weapons. One of the first questions asked of him was whether the detonation of an atomic bomb would ignite the Earth's atmosphere or oceans; physicist Edward Teller

An excerpt from Hans Bethe's notebook is part of the collections of the National Security Research Center.

Hans Bethe's personnel card documents details of his employment with the secret lab in Los Alamos. Initially reluctant, Bethe joined the Manhattan Project because he feared Nazi Germany was also working to develop nuclear weapons.

Hans Bethe, his colleague Enrico Fermi, and two children of scientists also employed at the Los Alamos lab. Nick King, sitting in the driver seat, later became a Los Alamos physicist himself.

In 1944, when the secret laboratory was creating an implosion type weapon, Bethe worked diligently to mathematically prove the feasibility of such an idea. His work continued as Bethe studied the hydrodynamic (fluid motion) aspects of implosion, which required him to determine how liquids and gases behaved at extreme temperatures and pressures. Bethe also assisted with the development of a neutron initiator, which is the trigger, or the fissile material that initially is spread out and is then pushed together to reach critical mass.

After the war's official end on September 2, 1945, Bethe returned to Cornell, but continued to be involved with the lab's development of a thermonuclear weapon, known as the hydrogen bomb or H-bomb. Bethe was uniquely adept for this project due to his breakthrough discovery in stellar nucleosynthesis. This is the process by which elements are created within stars by combining the protons and neutrons from the nuclei of lighter elements. Los Alamos scientists posited that the same type of energy generation that kept stars burning bright would lead to the successful detonation of a thermonuclear weapon.

suggested that detonation could lead to the destruction of the entire planet. Bethe theoretically proved the likelihood of such a catastrophe was vanishingly small and work on the weapon continued.

Bethe was responsible for completing other types of theoretical work that served as a basis for conceptualizing the first nuclear weapons. He calculated the critical mass, or the smallest amount of fissile material needed for a sustained nuclear reaction for a fission bomb like Little Boy, which was released above Hiroshima on August 7, 1945. Together with physicist and future Nobel laureate Richard Feynman (see Page 70), Bethe also developed a formula that estimated a bomb's explosive yield. This helped determine the effect of shock waves as well as the optimum altitude for detonation.

Bethe strongly opposed the development of the hydrogen bomb, but the Korean War in 1950 convinced him otherwise. Again, Bethe

Bethe was uniquely adept for this project due to his breakthrough discovery in stellar nucleosynthesis.

determined that if he "did not work on the bomb, somebody else would," according to the book *In the Shadow of the Bomb*, and he hoped his involvement with the new weapon would grant him the credibility and power to promote disarmament.

Before he agreed to participate in nuclear weapons work, Bethe recalled deeply contemplating the moral implications. Later in life he emerged as "the science community's liberal conscience," according to Bethe's obituary in *The New York Times*.

A proponent of arms control and nuclear disarmament, Bethe eventually sought to discontinue the creation of any new types of nuclear weapons. He long maintained that he never regretted his role in the creation of atomic bombs, viewing it instead as a difficult decision influenced by his fear of Nazism and a concern for the wellbeing of humanity.

His scientific research continued well into his 90s, with Bethe publishing at least one notable paper in the field of physics during every decade of his seven-decade career. Bethe died on March 6, 2005 at the age of 98, still shining brightly. He was survived by his wife, Rose, and their children Monica and Henry.

Hans Bethe's badge photo, issued after his arrival in Los Alamos on April 8, 1943.

HANS BETHE

1906–2005

LOS ALAMOS CONTRIBUTIONS

Theoretically proved the feasibility of Little Boy, Fat Man, and the hydrogen bomb. Calculated critical mass for Little Boy, which allowed scientists to successfully build the weapon. Developed the theory supporting the successful implosion method used at both the Trinity test and detonation of the Fat Man nuclear weapon.

NOBEL PRIZE: PHYSICS, 1967

Discovered the carbon-nitrogen-oxygen cycle, which demonstrates a series of nuclear reactions that produce enough stellar energy for the sun to burn for billions of years.

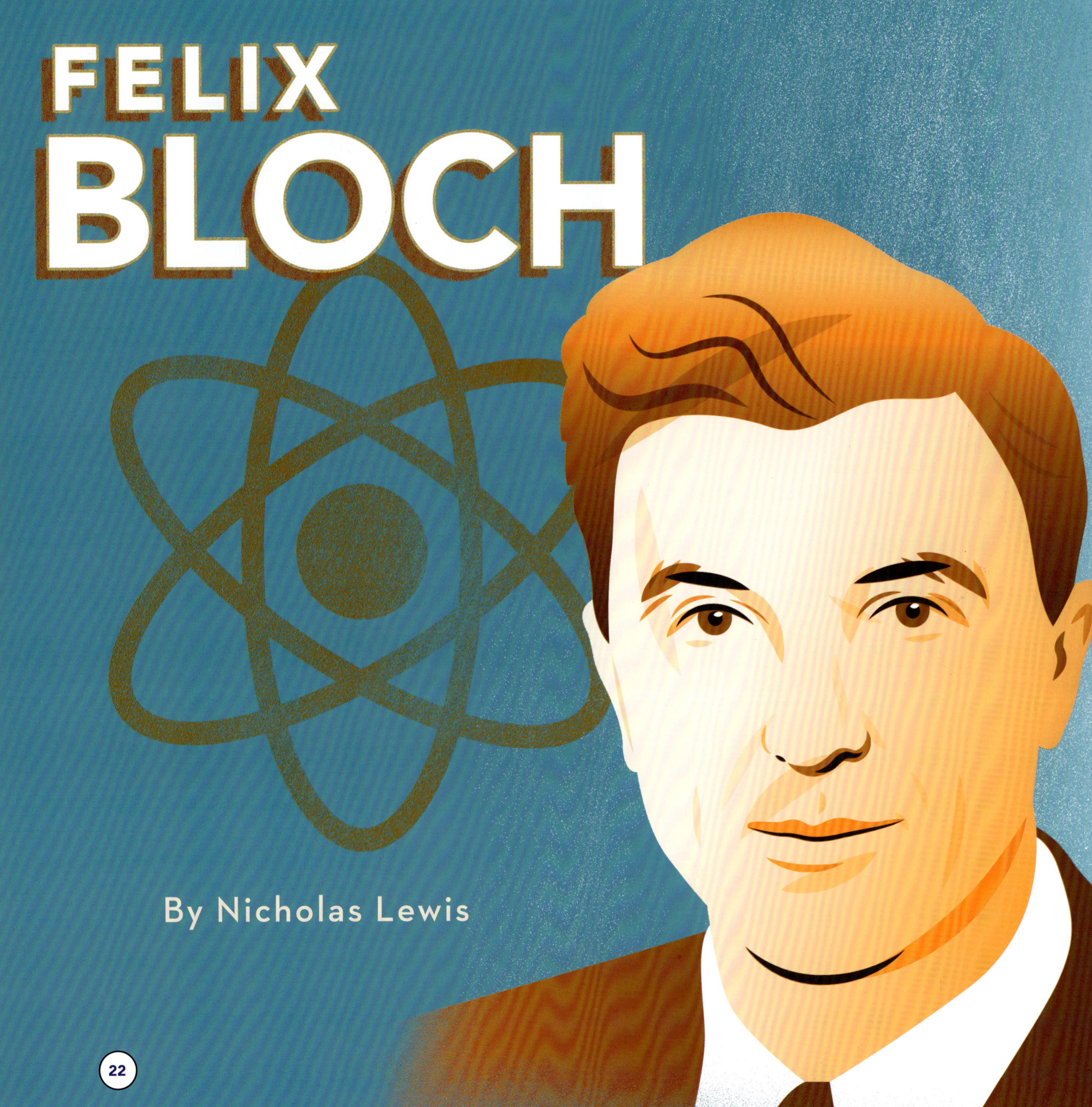

FELIX BLOCH

By Nicholas Lewis

"What I have achieved at Stanford (University)," Felix Bloch wrote to J. Robert Oppenheimer just months before joining the lab at Los Alamos in 1943, "would have been quite impossible without an everlasting readiness to dig into details and dirty work."

This was how Bloch, a Swiss-American physicist and former student of scientist and Nobel laureate Werner Heisenberg, explained his belief that the distinction between theoretical and experimental physics was an increasingly arbitrary one. The dirty work that Bloch described — combining theory and experiment — characterized his many pioneering contributions to science, his wartime work at Los Alamos, and his path to the Nobel Prize.

Early life and education

Born in Zurich in 1905, Felix disliked his early education until he encountered the clarity of mathematics and the beauty of music at 8 years old. He quickly became as adept at manipulating numbers as he was at playing Johann Sebastian Bach's harmonies on the piano. Encouraging their son's aptitude, his parents enrolled a 12-year-old Bloch into the renowned Canton of Zurich Gymnasium, an advanced secondary school that featured a rigorous curriculum in the sciences, mathematics, and languages. The school prepared students for Switzerland's challenging university placement exam, which Bloch passed with ease.

At the urging of his father, Bloch enrolled at Zurich's Federal Institute of Technology in 1924 to become an engineer. To his father's disappointment, Bloch's dislike of drafting and a fascination with theoretical work soon drew Bloch to physics. "I couldn't help it," he later explained.

In 1927, Bloch began his graduate education at the University of Leipzig in Germany, where he studied theoretical physics as Heisenberg's first graduate student — a partnership that would greatly influence Bloch's career. Heisenberg is credited for pioneering work in

Felix Bloch left Los Alamos in 1943 before the atomic devices were tested and deployed in combat in 1945, saying he could not live in the "military atmosphere" anymore.

quantum mechanics, which is the theoretical basis of modern physics that explains matter and energy on the atomic and subatomic level. At Heisenberg's suggestion, Bloch applied quantum mechanics (the mathematical description of motion and interaction of subatomic particles) to the classical theory problem of conductivity in crystal lattices, or the arrangement of atoms inside a crystal. This guidance influenced Bloch's 1928 dissertation topic, which developed the theory of conduction in metals and introduced the concept of "Bloch waves." This had enormous implications for semiconductor research.

Bloch then spent several years as a visiting scholar at multiple European centers of physics. He was supported by fellowships and assistantships, which allowed him to make many visits to physicist and Nobel laureate Niels Bohr (see Page 42) in Copenhagen, leading to a lifelong friendship.

During this period of continuous travel, Bloch conducted groundbreaking work to bridge the theory of electrical conductivity with experimental results. His work with Heisenberg on ferromagnetism (a characteristic of certain materials that can be magnetized) identified "Bloch walls," or the boundaries between magnetic domains.

Fleeing Nazi Germany and nuclear research

By 1932, Bloch had returned to Leipzig as a lecturer, but soon realized that he could not remain in Germany for long. Leipzig was a hotbed of anti-semitism during the rise of Nazism, and Bloch was a Jew.

Bloch fled Germany in 1933 for Rome, where he worked with physicist and Nobel laureate Enrico Fermi (see Page 64), followed by stints in Paris and Copenhagen. It was then that Bloch received a job offer from Stanford University, which sought scholars displaced by the Nazis. Knowing only that it was "somewhere on the West Coast," he left for the United States in 1934.

Bloch arrived at Stanford as an acting associate professor and was the only theorist in the physics department. He immediately collaborated with others in the field and, alongside future Los Alamos Lab Director J. Robert Oppenheimer, who was then at Berkeley, established a weekly theoretical physics seminar that became a popular event in the physics community. Encouraged by Bohr, Bloch researched the recently

discovered neutron, beginning a new line of neutron-scattering experiments inspired by his previous work on ferromagnets, or materials that can form permanent magnets, such as iron. In 1941, Bloch and his colleagues constructed a compact particle accelerator called a cyclotron, which propelled charged particles to extremely high speeds and energies. It quickly proved valuable for refining the neutron experiments, and would also prove useful for another, more urgent project.

Secret wartime work

As World War II raged, the Chicago Metallurgical Lab, working on behalf of the U.S. government, contracted with Bloch and his department to conduct classified research for the development of an atomic bomb. Using the cyclotron, Bloch determined the energy distribution of neutrons emitted during the fission process, which was research deemed so urgent that an extremely rare, 2-inch diameter sphere of uranium was shipped to Bloch's team with the request that they begin work before the contract was complete or security clearances were in place.

The cyclotron, a compact particle accelerator, was one of Bloch's most valuable contributions in the study of neutrons in the fission process.

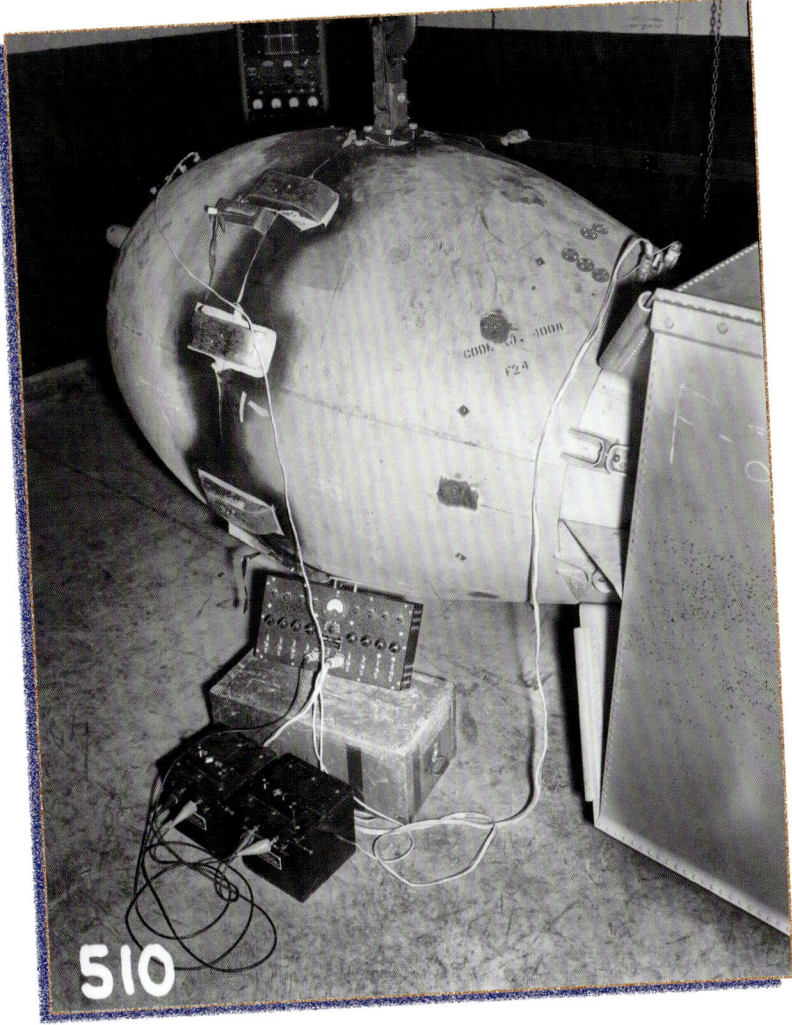

Oppenheimer deemed it worthy of further exploration. Bloch took part in Neddermeyer's experiments, later saying he had "very little doubt about it — that the velocities and pressures which one would get were indeed those which Neddermeyer said, and everyone said. But they needed experimental proof, and I did that."

The implosion-style atomic bomb was tested on July 16, 1945 in the New Mexico desert and a second bomb — codenamed Fat Man — was released just weeks later above Nagasaki, Japan. It was the second atomic weapon to be used in combat, following the uranium gun-type device named Little Boy.

Before atomic bombs were detonated, however, Bloch left Los Alamos. He later said, "I just could not live under this atmosphere. It was a military atmosphere. Letters were opened and one was under constant surveillance and so forth." The Blochs departed in November 1943, "somewhat to the annoyance of some of my friends, in particular Oppenheimer."

While at the Los Alamos lab, Felix Bloch focused on implosion experimentation, which led to the development of the Fat Man weapon.

Oppenheimer asked Bloch to join the newly formed, top-secret laboratory in New Mexico that would design and construct the atomic weapons. Oppenheimer wanted Bloch on staff in Los Alamos as quickly as possible, and readily agreed to Bloch's request for a financial bonus to move him, his wife, and their 2-year-old twin boys to New Mexico in the summer of 1943.

Bloch eventually joined physicist Seth Neddermeyer's Implosion Experimentation Group in the Ordnance Division. Neddermeyer was an early proponent of using conventional explosives to implode a subcritical mass of fissionable material into a supercritical state. This implosion concept did not initially receive widespread support at the Lab, but

The Fat Man plutonium implosion weapon was released above Nagasaki, Japan, on August 9, 1945. It was the second of two atomic bombs to be used in combat, helping to end World War II weeks later.

Felix Bloch's inprocessing card for employment at the lab includes Bloch's first, brief visit to Los Alamos in April 1943.

Bloch, Felix
From: Stanford
Salary: $500.00
Married: yes
m.2 yr.
Arrival: 4/22   to site
325 Le Conte Hall, Univ. of Calif., Berkeley
Mrs. Felix Bloch, 1551 Emerson, Palo Alto
wife Lore C. Bloch
Mrs. Bloch and 2 children to Site
7/10

A Nobel Prize

Bloch's work for the war, however, continued. He joined Harvard's Radio Research Laboratory to work on defenses against German radar. His work there had a significant influence on his post-war research, which would lead to the Nobel Prize.

He returned to Stanford in September 1945 and combined his wartime experience with microwaves and electronics work from Harvard with his previous research into ferromagnetism and magnetic moments. Bloch believed that his radio technology work at Harvard offered a simpler means of conducting measurements in nuclear moments.

The technique Bloch devised exploited the fact that protons and neutrons behaved like rotating magnets, causing atoms and molecules to align with magnetic fields. Exposing those atoms and molecules to radio waves would temporarily disturb their rotation. When they returned to their previous alignments, they would emit electromagnetic radio waves that characterized the different elements and isotopes in a compound.

In late 1945, after construction of the needed radio equipment, Bloch collaborated with Edmund Purcell, who had performed similar work, to demonstrate that nuclear magnetic resonance (NMR) was both simple and accurate, with profound implications for many fields. For physicists and chemists, NMR offered a completely electromagnetic and nondestructive means of studying the composition of different materials, and for medicine, it formed the basis of magnetic resonance imaging (MRI).

A musician, mountaineer, and lover of art, Bloch was also known for his support of humanitarian causes, including advocating for colleagues detained in the Soviet Union.

Bloch and Purcell were awarded the 1952 Nobel Prize in Physics for their development of new methods for nuclear magnetic precision measurements.

In 1954, Bloch became the first director of CERN, the European Organization for Nuclear Research, then spent the remainder of his career pursuing theoretical and experimental physics at Stanford. A musician, mountaineer, and lover of art, Bloch was also known for his support of humanitarian causes, including advocating for colleagues detained in the Soviet Union.

Following Bloch's retirement, the family visited Bloch's hometown of Zurich, where he died of heart failure in September 1983. He was survived by his wife, daughter Ruth, and sons Daniel, Frank, and George.

Felix Bloch is pictured here in the 1950s with a nuclear magnetic resonance (NMR) spectrometer. The NMR was a completely electromagnetic means of studying the structural composition of different materials, and led to Bloch's Nobel Prize in Physics. (Public domain image.)

FELIX BLOCH

1905–1983

LOS ALAMOS CONTRIBUTIONS

Played a key role in early experiments that helped to demonstrate the viability of an implosion weapon.

NOBEL PRIZE: PHYSICS, 1952

Developed the nuclear magnetic resonance technique for studying the composition of materials.

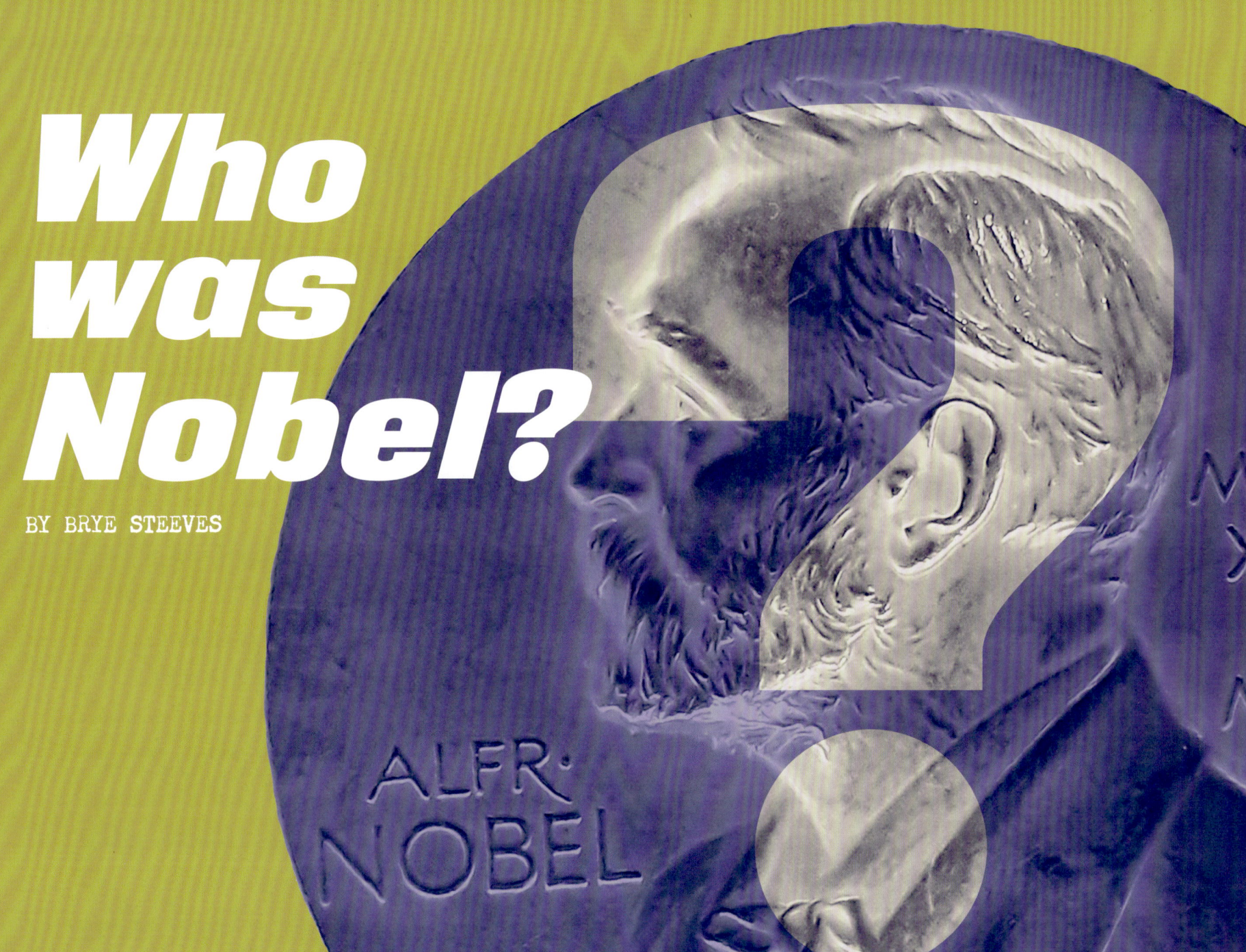

Who was Nobel?

BY BRYE STEEVES

ALFR· NOBEL

(ID 176794226 © Robert Paul Van Beets | Dreamstime.com.)

Eight years before his death, Alfred Nobel read his own obituary.

And, in addition to his demise, of course, he didn't like what he saw.

It was 1888 when one of Nobel's brothers died and a newspaper mistakenly published an obituary about Alfred. The article called him "the merchant of death," referring to the significant fortune he had amassed in the explosives and munitions businesses. Most notably among the 355 patents Nobel held was dynamite, which he invented in 1867. He thought it would put an end to war, but instead dynamite was widely viewed as an extremely deadly product rather than as a deterrent to large-scale conflict.

Biographer Kenne Fant wrote that Nobel "became so obsessed with his posthumous reputation that he rewrote his last will, bequeathing most of his fortune to a cause upon which no future obituary writer would be able to cast aspersions."

Nobel never spoke publicly about the motivations behind establishing the Nobel Prize. Once his plans for the prize were finalized, he established annual prizes in five categories: physics, chemistry, medicine, literature, and peace. (A sixth category to recognize achievements in economics would be added in 1968 by the Bank of Sweden.)

The Nobel Prizes were to be bestowed upon "those who, during the preceding year, shall have conferred the greatest benefit on mankind." Each year, Nobel's wealth (equivalent to approximately $265 million in today's dollars adjusted for inflation), would be awarded to individuals or organizations, along with a large monetary prize and a gold medal.

Nobel never married or had children, and upon his death from a brain hemorrhage at age 63 in 1896, many of his extended family members were surprised by his will and wanted to contest it. Ultimately, his wishes were upheld, but because many details in the will were incomplete, the Nobel Foundation was not established and the first Nobel Prizes were not awarded until the fifth anniversary of his death on December 10, 1901.

The prizes have been awarded annually since then. The ceremonies take place in Nobel's native Stockholm, Sweden, with the exception of the Nobel Peace Prize, which is presented in Oslo, Norway. More than 960 people and 25 organizations have received a Nobel Prize. Four people have received more than one Nobel Prize.

The Nobel Prize categories — and recipients, called "laureates" — are perhaps a representation of his life's interests. Although a scientist by trade and a successful entrepreneur, Nobel also loved literature and wrote poetry and drama. He was fluent in Swedish, Russian, French, English, and German.

Richard Feynman (see Page 70) receiving the Nobel Prize in Physics in 1965. (Photo courtesy of the Archives, California Institute of Technology.)

AAGE
BOHR

By Ellen D. McGehee

When 21-year-old Aage Bohr arrived at Los Alamos on

December 30, 1943, he was one of the youngest members of the

laboratory's scientific staff. Working with his father, Niels Bohr, Aage would

make significant contributions in support of the war effort.

Early years

Aage Niels Bohr was born on June 19, 1922, in Copenhagen. As a young man, Bohr's university studies were interrupted by the German occupation of Denmark. After escaping to Sweden with his family in October 1943, Aage joined his physicist father, Niels Bohr, in England, and the two arrived in the United States at the end of November of the same year.

Los Alamos and the Manhattan Project

Aage Bohr, like his father, was a member of the Manhattan Project's British Mission, a formal team of British scientists stationed at wartime Los Alamos.

At Los Alamos, codenamed Project Y, Aage and his father were accorded VIP status because of the family's scientific renown and were given new names for the duration of the war to hide their identities. Aage Bohr was known as James Baker or just Jim. His father was given the name Nicholas Baker on official documents, but the Project Y scientific community knew him as Uncle Nick.

During the war years, Aage Bohr also functioned as a personal secretary to his father, but he was much more than that. As Lab Director J. Robert Oppenheimer later noted in 1964, Aage was Niels Bohr's "companion, his confidante."

Aage's role as a companion to his father was noted in many post-war accounts. Physicist Enrico Fermi's wife, Laura, wrote about the Bohrs' relationship in her memoir, *Atoms in the Family*, stating that "when he [Niels Bohr] walked about town, he did not seem to care where he was going. He let himself be led by his young son, a physicist like himself, who never left his side."

Fuller Lodge was a frequent dining spot for Los Alamos residents and lab staff, including physicist Aage Bohr and his father, fellow physicist Niels Bohr.

Aage Bohr often served as an intermediary between his father and other scientific staff, even during social occasions. Physicist Hugh Richards, writing in a memoir account, describes how Niels liked to tell jokes at dinner in Fuller Lodge. Aage would save their table guests the awkwardness of not understanding his father's heavy accent: "Fortunately, Aage always laughed at his father's jokes so even when we missed the point, we got along OK by cueing our laughter to Aage's." During their visits to Los Alamos, Aage Bohr and his father took daily walks, which were detailed by their assigned security detachment:

Both the father and son appear to be extremely absent-minded individuals, engrossed in themselves, and go about paying little attention to any external influences. As they did a great deal of walking, this Agent had occasion to spend considerable time behind them and observe that it was rare when either of them paid much attention to stop lights or signs, but proceeded on their way much the same as if they were walking in the woods.

A February 9, 1945, memo from lab Director J. Robert Oppenheimer announcing the formation of an advisory committee on implosion initiators. The 1945 Oppenheimer memo is part of the collections in the National Security Research Center (NSRC), the lab's classified library.

UNCLASSIFIED CLASSIFICATION CANCELLED PER DOC REVIEW JAN. 1973

VERIFIED UNCLASSIFIED
NOV - 3 1980

February 9, 1945

FINAL DETERMINATION
UNCLASSIFIED
L. M. Redman
NOV 3, 1980

TO: H. A. Bethe
R. F. Christy
E. Fermi

SUBJECT: Advisory Committee on Design and Development of Implosion Initiators

1. I should like to have you serve as an advisory committee on the design and development of the implosion initiators.

2. Dr. N. Baker has signified his willingness to meet with the committee during his visits to Y, but regards it as inappropriate to accept formal membership on the committee.

3. I believe it important that the committee maintain close contact with Dr. Critchfield and Dr. Bacher, who are in immediate charge of the work, and feel free to have them attend its meetings. The committee should, however, not hesitate to recommend that other parts of the laboratory be asked to contribute to the work if in its opinion this is desirable.

4. The committee should also feel free to meet with Dr. Dodson and other members of his group on all questions concerning the handling and preparation of polonium.

5. I should like to have Dr. Bethe serve as chairman of this committee

J. R. Oppenheimer

UNCLASSIFIED

cc: Dr. Bacher
Dr. xxxxx Baker
Dr. Critchfield
Dr. Dodson ✓

CLASSIFICATION CANCELLED
PER DOC REVIEW JAN. 1973

But it was during these regular walks that Aage Bohr provided perhaps his most important service to Los Alamos and the Manhattan Project. Writing in his memoir about wartime Los Alamos, scientist Joseph Hirschfelder observes that "Bohr and his son, Aage ... were almost inseparable. Each day they would take a long walk during which they would discuss some very difficult physics problem." By functioning as his father's scientific sounding board, Aage Bohr helped solve some of the most pressing theoretical problems facing Project Y's scientists.

The initiator dilemma: using theory to prove practice

One of the biggest challenges faced by the wartime laboratory in its quest for deliverable weapons designs was the development of a proven initiator device, which is located inside the weapon and helps trigger the nuclear reaction. This challenge was made more difficult because the initiator's performance could only be verified in an actual weapon test. Preliminary experiments were set up at a test range at Sandia site, located along the lower section of present-day East Jemez Road in Los Alamos. Later initiator experiments were conducted at P-Site, a wartime technical area located within the boundaries of today's S-Site. Although the Initiator Group was tasked with developing and testing various designs, the lack of experimental evidence resulted in the creation of a special committee in February 1945, of which Niels Bohr was a member.

Compounding the issue, Enrico Fermi was skeptical of the preferred initiator model known as the urchin. Presumably unbiased as relative outsiders, theoreticians Niels and Aage Bohr were asked to review the initiator data. After several days of discussion, the Bohrs independently vetted the urchin design, which was subsequently approved for use on May 1, 1945.

Aerial photograph of wartime technical area P-Site (TA-13), the location of x-ray experiments conducted in 1945 for the initiator program.

Aage Bohr was a physicist like his father, Niels, and accompanied him to the Los Alamos lab to work. He was his father's constant companion, always by his side. (Photo courtesy of the Niels Bohr Archive.)

They are just one of four father-and-son sets to receive the physics prize.

In a 1984 interview, physicist Robert Bacher recalled the Bohrs' assistance with the initiator problem:

I said, "What I'd like to do is, Uncle Nick [Niels Bohr] is here now, and I'd like to go and explain to him about the initiator and say I'd like his advice and counsel on whether he thinks it will work or not."… So I talked to him for a long time and then he spent about two days with his son Aage going over every single thing that had been done on this business. I saw him after this and he said, "My, that's very impressive. I think that will work."

And it did. First at the Trinity test on July 16, 1945, when The Gadget's successful detonation marked the dawn of the Atomic Age, and then weeks later in combat when Fat Man was released above Nagasaki, Japan, on August 9, 1945, helping to end World War II shortly thereafter.

Post-war and a Nobel Prize

Aage Bohr returned to Denmark in the summer of 1945 and received a master's degree in physics from the University of Copenhagen in 1946. He spent time in the United States from 1948 to 1950, where he continued his studies at Princeton and Columbia universities. Completing his Ph.D. at the University of Copenhagen in 1954, Aage Bohr spent most of his academic and professional career in Denmark. In the early 1960s, after a stint as a university professor, he took over the directorship of the Niels Bohr Institute, founded by his father and, from 1975 to 1981, was the director of the Nordic Institute for Theoretical Atomic Physics.

Aage Bohr was awarded the Nobel Prize in Physics in 1975, 53 years after his father, who also received the prize (see Page 42). They are just one of four father-and-son sets to receive the physics prize.

Aage Bohr's prize-winning research on the structure of the atomic nucleus was conducted many years earlier while working with fellow physicist Ben Mottelson at Columbia University, prior to receiving his doctorate from Copenhagen. The prize was awarded in equal shares to Bohr, Mottelson, and James Rainwater of the United States, who had independently proposed what the Nobel committee described as the "connection between collective motion and particle motion in atomic nuclei." This connection led to a new theory of atomic structure, in which collective motion results in a deformation of the shape of the nucleus. On September 8, 2009, Aage Bohr died in Copenhagen at the age of 87. He was survived by his wife, Marietta, and their sons, Thomas and Vilhelm.

AAGE BOHR

1922–2009

LOS ALAMOS CONTRIBUTIONS

Junior scientific officer who visited Los Alamos several times between 1943 and 1945 with his father, Niels Bohr, as a member of the British Mission. Aage Bohr was a consultant to the Manhattan Project and, working alongside his father, contributed to the theoretical confirmation of the initiator concept, which helps trigger the nuclear reaction, for the Fat Man weapon design.

NOBEL PRIZE: PHYSICS, 1975

Expanded the scientific understanding of the shape of the nucleus by researching the rotational motion of protons and neutrons. This research revealed asymmetrical changes to the nucleus; previous models had suggested that nuclei were always spherically shaped.

NIELS BOHR

By Alan B. Carr

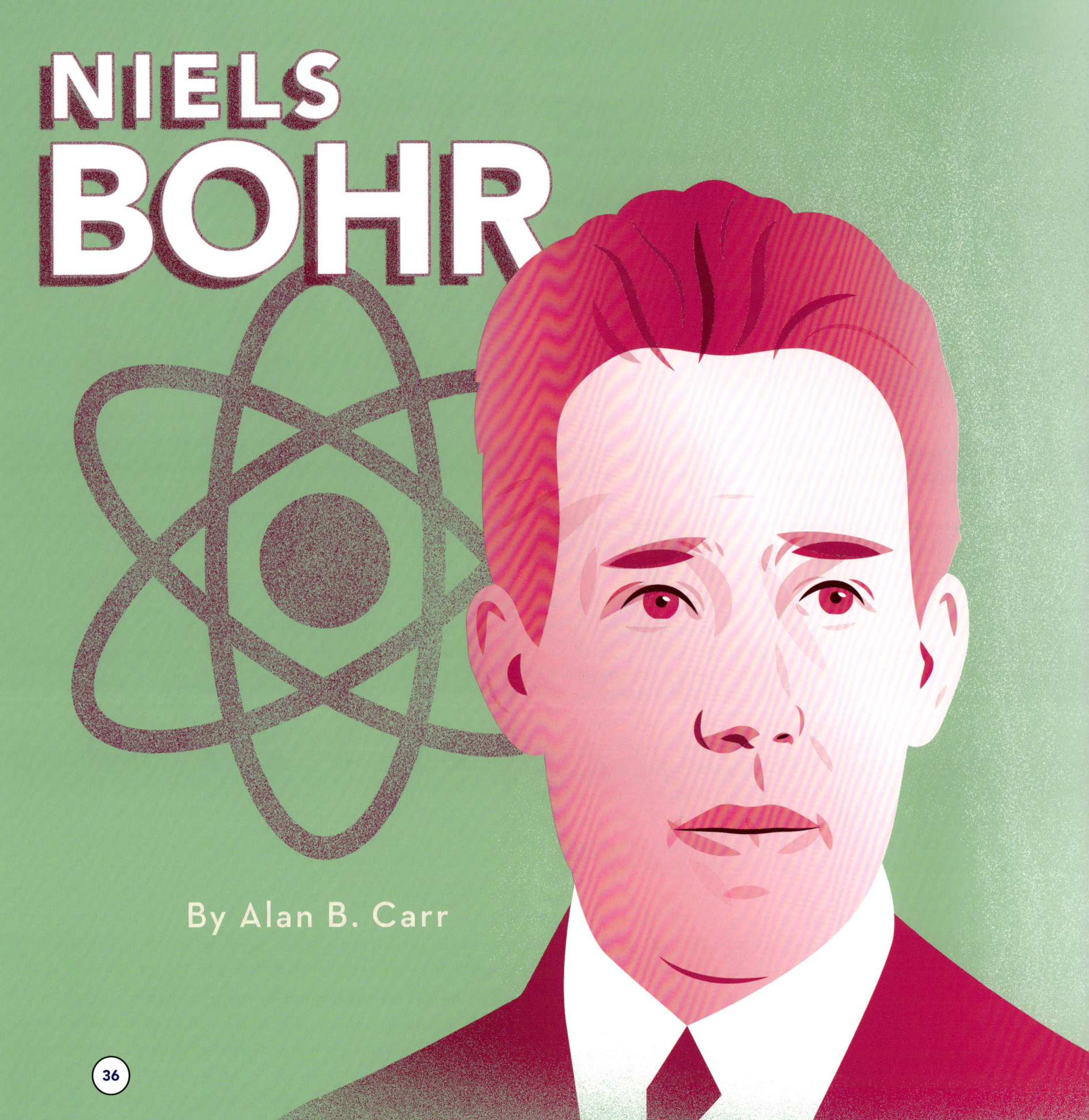

At the end of September 1943, Niels Bohr found himself aboard a boat bound for Sweden.

Behind him were Denmark, his homeland, and certain arrest by the Gestapo. Ahead:

opportunities to shape a more peaceful future for the world. But at that moment in time, the light of

the rising moon chipped away at the precious darkness concealing his escape. This is just one of the

harrowing events in the life of one of history's most interesting and respected scientists.

Early years

Ellen Bohr gave birth to her middle child, Niels, in Copenhagen on October 7, 1885. Her husband Christian was a prominent physiologist, thrice nominated for — though never awarded — the Nobel Prize in Physiology or Medicine. As children, Niels and his younger brother Harald, who would become a well-known mathematician, listened to their father's conversations with academics, which stimulated their interest in the natural sciences from an early age.

Bohr attended the University of Copenhagen and received his Ph.D. in physics in 1911, though his advisor conceded there was no one in Denmark knowledgeable enough on the electron theory of metals (that a metal contains a gas of electrons that are completely free to move within it) to properly evaluate his work.

The next year, Bohr married Margrethe Nørlund, the sister of one of Harald's students. The happy marriage lasted more than half a century and produced six sons: Christian, Aage, Ernest, Hans Henrik, Erik, and Harald.

In fall 1914, after World War I started, Bohr left neutral Denmark to work in Manchester, England at the laboratory of British physicist and 1908 Nobel laureate Ernest Rutherford. There, he lectured and published papers that garnered international renown. By 1916, a professorship in theoretical physics was established for Bohr back at Copenhagen University. Over the years, Bohr published groundbreaking papers on quantum theory and secured government

and private support for his Institute of Physics, which opened in March 1921. Three years after his death in 1962, it was renamed the Niels Bohr Institute.

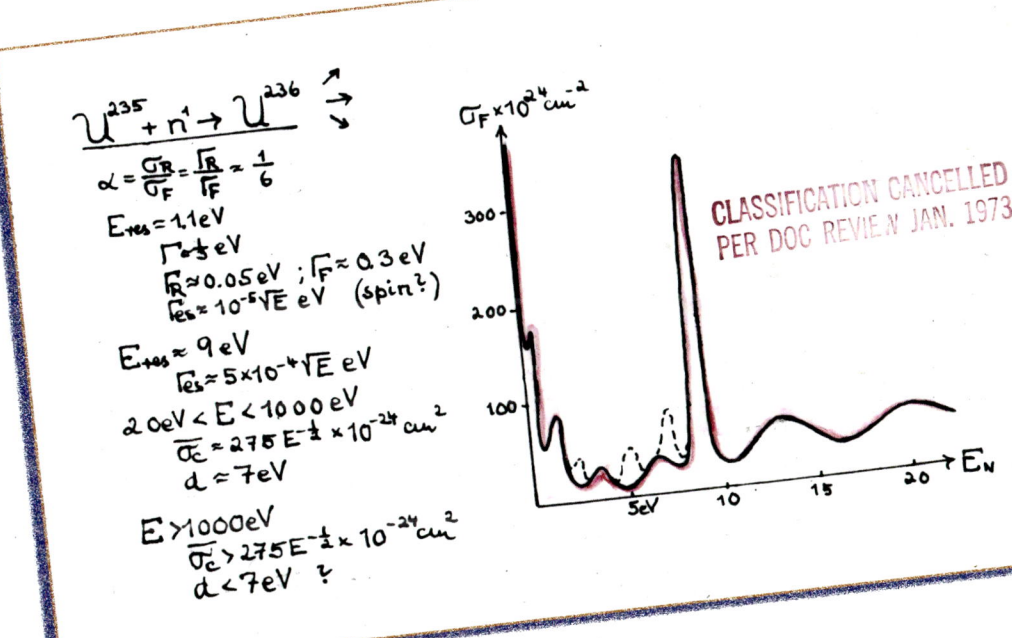

An excerpt from one of Niels Bohr's slides from his colloquium presentation on November 28, 1944 is shown here. Colloquiums were held often during the Manhattan Project for the benefit of the scientists.

A Nobel Prize and debates with Einstein

Fellow Danish physicist Rud Nielson remembers Bohr constantly on the move. The energetic Bohr "was always friendly and less remote and dignified than most Danish professors in those days." Known for his quiet, difficult-to-understand voice and gentle nature, Bohr was awarded the Nobel Prize in Physics in 1922 "for his services in the investigation of the structure of atoms and of the radiation emanating from them." Understanding the structure of atoms came as the result of many significant discoveries. Bohr's theory of atomic structure was revolutionary at the time because it incorporated concepts from German physicist and 1918 Nobel laureate Max Planck's quantum theory into Rutherford's proposed atomic model. Bohr's theory succeeded in modeling experimental observations in hydrogen emission spectra (frequencies of electromagnetic radiation emitted due to the transition from a high energy state to a lower energy state) and his work was later refined with the development of quantum mechanics.

As the 1920s progressed, Bohr's emerging ideas on quantum mechanics, which examine the behavior of matter and light on the atomic and subatomic scale, came in conflict with those of Albert Einstein. The two legendary scientists regularly engaged in friendly debates over the nature of physical reality. Observations of paradoxes in nature prompted Einstein to famously state, "God does not play at dice." Lesser known is Bohr's response: "But still, it cannot be for us to tell God, how he is to run the world."

The Copenhagen interpretation of quantum mechanics emerged, in part, as a result of these exchanges. Primarily conceived by Bohr, this approach embraces complementarity: that physical entities possess various attributes that cannot be measured at the same time. Thus, the same object can seem very different depending on how, or if, it is observed. Nearly a hundred years after it was first proposed, complementarity remains a foundational concept of quantum mechanics.

Some of Niels Bohr's notes are part of the collections of the National Security Research Center, which is the classified library at Los Alamos National Laboratory. The NSRC also houses unclassified legacy materials.

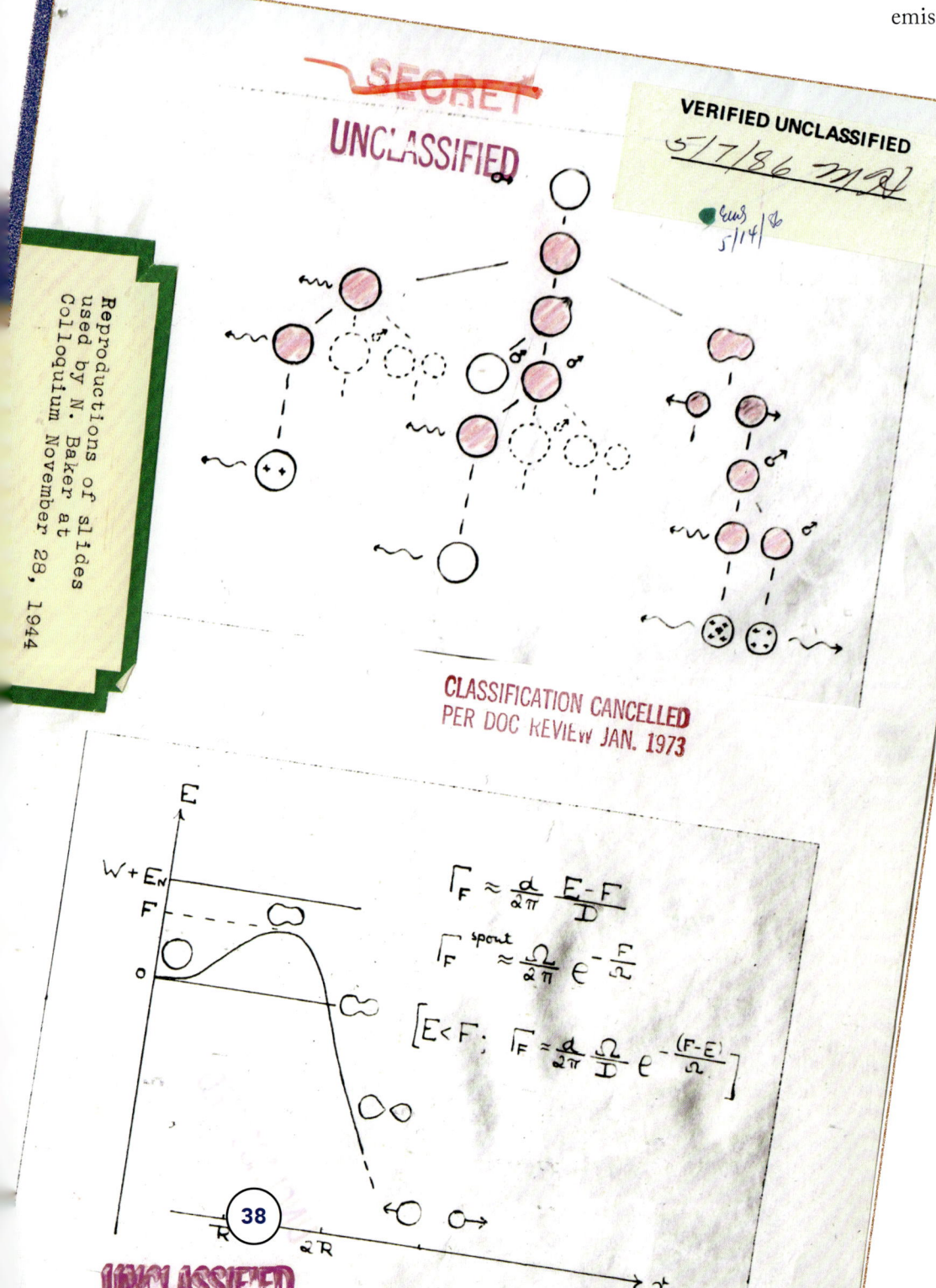

Reproductions of slides used by N. Baker at Colloquium November 28, 1944

VERIFIED UNCLASSIFIED
5/7/86

UNCLASSIFIED

CLASSIFICATION CANCELLED PER DOC REVIEW JAN. 1973

UNCLASSIFIED

Fission and Nazis in Germany

The unknowing production of fission (a reaction in which the nucleus of an atom splits into two or more nuclei) in Nazi Germany in late 1938 and the realization of its discovery in early 1939 quickly led scientists to speculate whether the process could be harnessed to create a bomb. Tragically, 1939 would also witness the beginning of World War II.

In October 1941, Bohr had a fateful meeting in Copenhagen with longtime colleague and German physicist Werner Heisenberg. Heisenberg shared information with Bohr on nuclear weapons research in Germany. The exact nature of this meeting remains debated, but Bohr, who had given little serious thought to nuclear weapons design, found Heisenberg's report disturbing. By late summer of 1943, German soldiers occupied Denmark and the time had come for Bohr, who was Jewish, to flee.

The night of September 29, Bohr and his wife and children escaped to neutral Sweden by boat, and then a British plane secretly carried them to the United Kingdom. Bohr doubted the feasibility of nuclear weapons, but found himself in the midst of the secret Allied effort to build them.

The United States had taken the lead in weapons development and after spending about two months in Britain, Niels and his son Aage arrived in the United States to participate in the Manhattan Project, which was the U.S. government's top-secret effort to create the first-ever nuclear weapons.

On December 30 they arrived at the project's weapons-design laboratory in Los Alamos, New Mexico. For security purposes, the famous physicist and his son were given the codenames Nicholas and James Baker. At Los Alamos, the duo became affectionately known as "Uncle Nick" and "Jim." Aage, a future Nobel laureate (see Page 36), served as his father's ever-present secretary. The two did not permanently join the Los Alamos staff but, rather, acted as key consultants who made regular trips to the laboratory.

Los Alamos contributions

If not critical in nature, Bohr's technical contributions at Los Alamos remain notable. For instance, he contributed to the design of the implosion weapon's initiator: a neutron-generating component designed to help start the nuclear chain reaction. Bohr sat on the Initiator Committee, which included physicist Robert Christy, future Nobel laureate Hans Bethe (see Page 22), and 1938 Nobel laureate Enrico Fermi (see Page 64). Fermi lacked confidence in the concept, but Uncle Nick and Jim assured him that the initiator would likely work. In fact, Nicholas Baker is listed as an inventor on patent S-4344, the Modulated Neutron Source.

Bohr also gave some consideration to critical mass, which is the minimum amount of fissile material necessary for a nuclear detonation. In June 1944, while away from the laboratory, Bohr wrote to Director J. Robert Oppenheimer, who was in Los Alamos: "…it has occurred to me that the number of neutrons per fission may be considerably larger for fast neutrons than for slow ones, and that accordingly the critical mass of a fast working gadget may be essentially smaller than hitherto estimated." Bohr's enthusiasm for

He played a small part in developing nuclear weapons, but played an enormous role in attempting to harness the destructive nature of these devices for peace.

the design work was palpable: "I am very eager to discuss all such problems with you as soon as possible."

As the Allied victory came into view, Bohr met separately with Prime Minister Winston Churchill and President Franklin Roosevelt, hoping to persuade the two leaders to avoid a future nuclear arms race, but his warnings went unheeded. Bohr explained: "That is why I went to America. They didn't need my help in making the atom bomb." Nonetheless, Bohr returned to Los Alamos and held a symposium on November 28.

In addition to his technical contributions, Bohr inspired and challenged young staff to contemplate the potential consequences of the forthcoming Atomic Age. Robert Wilson, a group leader in the Physics Division, recalls: "Mostly I remember sitting on the floor in a group of awestruck listeners as Bohr agonized over the kind of world that would result from our grim work on nuclear energy. Actually, much of the conversation was of a playful nature; it scintillated." But the Trinity test of July 16, 1945 ended the playfulness: "As the message of the bomb's actuality sank in deeper and deeper, it became clear that I — that we all — shared the kind of responsibility that Bohr had so perceptibly been preaching about."

Optimism and science

After the collapse of Nazi Germany, Bohr returned to England in June 1945. In reflecting on Bohr's time at Los Alamos, Oppenheimer stated: "Bohr made it seem hopeful. He spoke with contempt of Hitler who … had tried to enslave Europe for a long, long time. He said nothing like that would ever happen again." With Hitler dead, Bohr went back to Denmark in August; that same month, nuclear weapons were unleashed against the Japanese cities of Hiroshima and Nagasaki.

After the war, Bohr concentrated on finding common ground between the Soviet Union and its former allies, the United Kingdom and the United States. He recognized that while atomic bombs made future nuclear wars possible, the devastating nature of the technology might help world leaders embrace openness and cooperation.

During the 1950s, Bohr spent less time in the laboratory and more time convincing governments to build new ones. Bohr supported the founding of the European Organization for Nuclear Research (CERN) and led in the creation of the Nordic Institute of Theoretical Physics. At the mandatory retirement age of 70 at the University of Copenhagen in 1955, Bohr bequeathed his position to his son Aage, but continued to lead the Institute of Physics. He chaired the newly created Danish Atomic Energy Commission and led the effort to build Denmark's national laboratory at Risø.

A lasting legacy

Bohr inevitably encountered opposition as he struggled for global peace and security for the world's residents, but he never encountered an enemy. Aage remembers, "In his attitude to political problems, as in all other matters, any differentiation between 'opponents' and 'supporters' was foreign to him, and he had a particular gift for entering into the spirit of the various points of view, and with this background working for a cause which he was convinced was in the deepest interest of all."

Bohr died of heart failure on November 18, 1962 at age 77. He played a small part in developing nuclear weapons, but played an enormous role in attempting to harness the destructive nature of these devices for peace. Bohr's ideas changed history, and even those ideas that have yet to be realized, such as his concept for an open world, continue to stimulate hope for a more peaceful future.

Niels Bohr's legacy centered on his conviction that although atomic bombs made future nuclear wars possible, they might also help world leaders embrace openness and cooperation. (Photo courtesy of AIP Emilio Segrè Visual Archives, Margrethe Bohr Collection.)

NIELS BOHR

1885–1962

LOS ALAMOS CONTRIBUTIONS

Contributed to the design of the implosion weapon's initiator, which was a neutron-generating component designed to help start the nuclear chain reaction.

NOBEL PRIZE: PHYSICS, 1922

Investigation of the structure of atoms and the radiation they release.

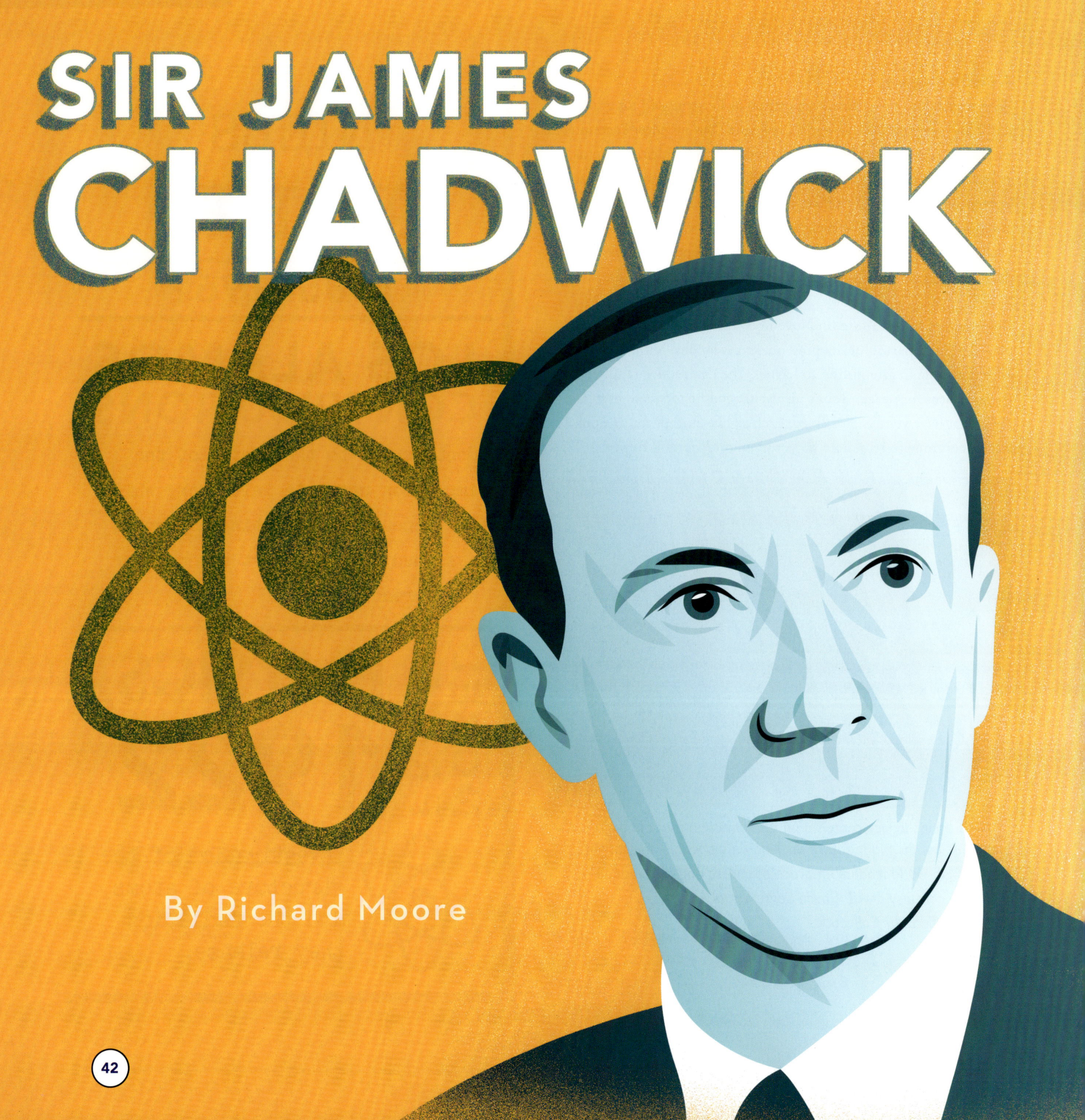

SIR JAMES
CHADWICK

By Richard Moore

The discoverer of the neutron was a Nobel laureate long before he arrived in

Los Alamos in 1944, and a knight before he left. In between these two honors,

James Chadwick wrote the reports that prompted the United States to begin in earnest the

development of the atomic bomb. His diplomacy, pragmatic advice, and tireless work brought

the U.S. and British atomic programs together to help end World War II.

Academics, honors, and a family

Chadwick came from a humble background. He was born near the industrial city of Manchester, England, in 1891. At just 16 years old, he won a scholarship to Manchester University, where he began a long association with Ernest Rutherford, who was known as the father of nuclear physics.

Chadwick earned his master's degree in science in 1912 and, just before the First World War began in July 1914, Rutherford recommended him for another scholarship. This included working with German physicist Hans Geiger in Berlin for one year. This turned into five years when Chadwick was interned as an enemy alien at a hastily constructed civilian detention camp at the Ruhleben racecourse in the Berlin suburbs. Improbably, Chadwick and other British internees were able to continue their scientific research, experimenting, for example, with the weak radioactivity in smuggled toothpaste. The cold and malnourishment, however, permanently affected Chadwick's health. He was released following the armistice with Germany in November 1918.

Back in Britain, Chadwick followed his mentor to the Cavendish Laboratory at the University of Cambridge, where, in time, he effectively became Rutherford's deputy. A quiet, reserved man, Chadwick preferred practical experimental work to teaching, but he made many contacts among the students and visitors drawn to the Cavendish, which was then the world's leading center for nuclear physics. Some of these contacts included Hans Bethe, Ernest

Pictured here at the 1933 Solvay physics conference in Brussels, James Chadwick is seated at the far right of this group of renowned scientists. Not long after, Chadwick would be awarded the 1935 Nobel Prize in Physics for his discovery of the neutron (Photo courtesy of Wikimedia Commons.)

Lawrence, and Mark Oliphant, all of whom he would meet again through work on the Manhattan Project.

In August 1925, the 34-year-old Chadwick married Aileen Stewart-Brown, and in February 1927 the couple had twin daughters, Judith and Joanna. Also that year, Chadwick became a fellow of the Royal Society in London, the United Kingdom's National Academy of Sciences.

By 1931, teams in France and Germany were conducting experiments with polonium and beryllium and were observing radiation with new and unexpected properties. Using a quick and small-scale experiment with a paraffin wax block, an ionization chamber, amplifier, and a photographic recorder, Chadwick showed this radiation to be an electrically neutral subatomic particle that he and Rutherford had previously predicted: the neutron. For his discovery of the neutron, Chadwick was awarded the Nobel Prize in Physics in 1935.

```
Chadwick, James
wife Aileen Chadwick
Arrival: 1/7/44
Judith and Joanna arrived 4/25/44

7/12/45
```

James Chadwick's major contribution to the Manhattan Project was in diplomacy between these two partner countries in their race against Nazi Germany to develop the atomic bomb.

Extensive preparations were made in the New Mexico desert for the Trinity test, conducted on July 16, 1945. James Chadwick was among the scientists, military personnel, and other observers to witness the dawn of the Atomic Age.

Chadwick left Cambridge for a chair position at Liverpool University in his wife's hometown, where there was funding and space to build a cyclotron — the very latest particle accelerator, now seen by many as the badge of a modern, forward-looking physics laboratory.

A superbomb and a mission to the U.S.

As one of the leading physicists in the United Kingdom, Chadwick was asked by the government in October 1939 about the feasibility of

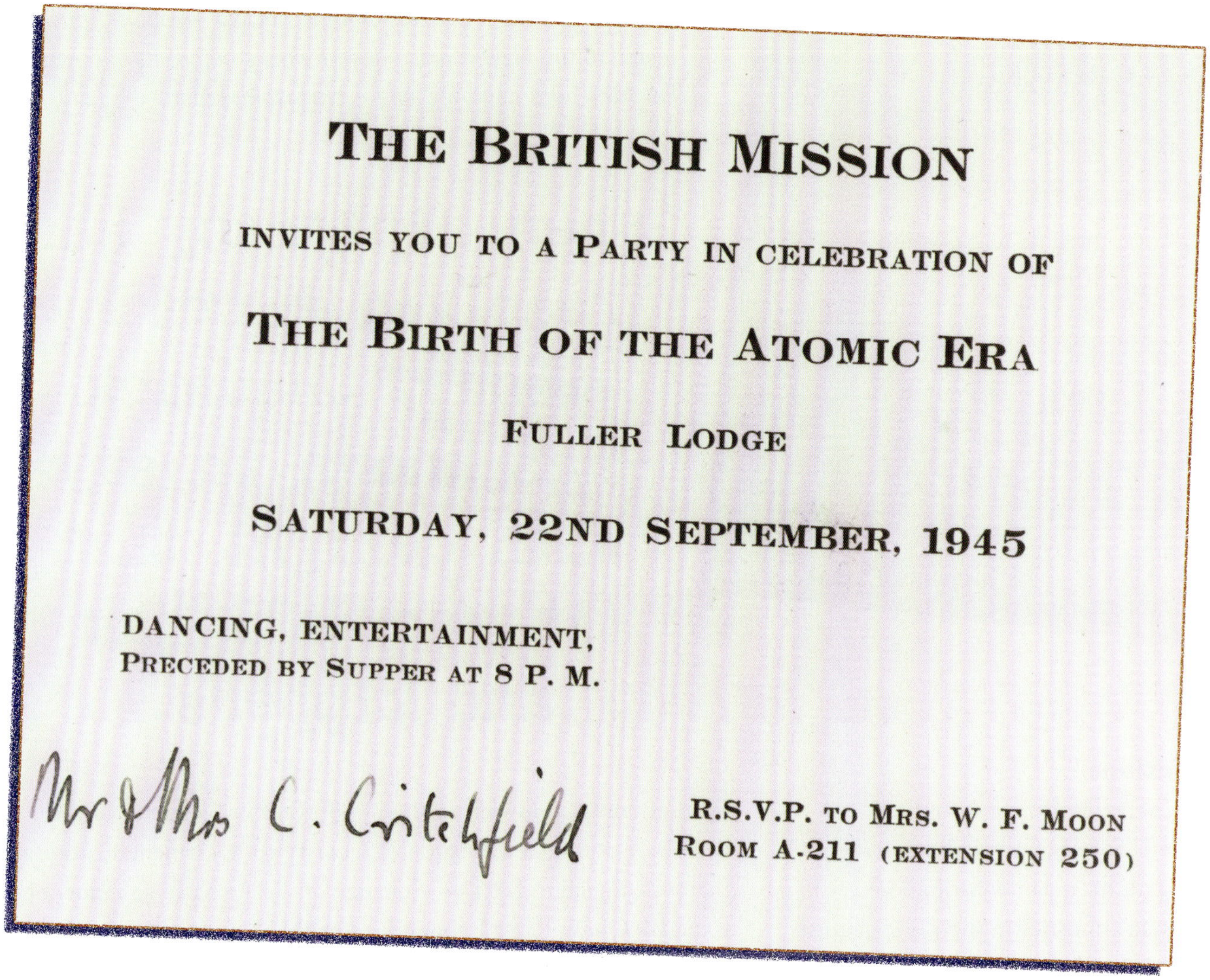

THE BRITISH MISSION

INVITES YOU TO A PARTY IN CELEBRATION OF

THE BIRTH OF THE ATOMIC ERA

FULLER LODGE

SATURDAY, 22ND SEPTEMBER, 1945

DANCING, ENTERTAINMENT,
PRECEDED BY SUPPER AT 8 P. M.

Mr & Mrs C. Critchfield

R.S.V.P. TO MRS. W. F. MOON
ROOM A-211 (EXTENSION 250)

The British Mission was a group of some of Europe's best scientists who worked alongside their American counterparts in support of the Manhattan Project. After the war officially ended on September 2, 1945, members of the British Mission in Los Alamos hosted a celebratory party that included a skit, dinner, and dancing.

an atomic bomb. Wanting more data before committing himself to a view, he set up experiments and discussed the idea with one of his colleagues, a young Polish refugee named Joseph Rotblat — a future Manhattan Project scientist in Los Alamos and future Nobel laureate (see Page 130).

Chadwick was not surprised, therefore, when in March 1940 Otto Frisch and Rudolf Peierls, two refugee scientists working at the United Kingdom's University of Birmingham, produced a memorandum outlining a workable "superbomb." In the summer of 1941, Chadwick personally wrote the official reports for the British scientific working group known as the MAUD Committee. The reports confirmed these findings and brought them to the wider attention of the British and U.S. governments.

When Prime Minister Winston Churchill and President Franklin Roosevelt agreed in August 1943 to a joint Anglo-American atomic project, Chadwick was chosen as the head of the British atomic energy mission to the United States. He spent most of the next two years in Washington, D.C., and also made several extended visits to Los Alamos, where his wife and daughters lived for much of 1944.

The Gadget was an implosion-type nuclear weapon that was tested in the New Mexico desert weeks before its successor, Fat Man, was deployed above Nagasaki, Japan.

When diplomacy had resolved a tricky issue, Chadwick was known to say with delight that it was "all jam and kippers."

Manhattan Project contributions

Chadwick did not contribute original scientific work to the Manhattan Project, which was the U.S.-led effort to develop the world's first nuclear weapons. Instead, his major contribution was in diplomacy between these two partner countries in their race against Nazi Germany to develop the atomic bomb.

In particular, Chadwick forged a close working relationship, even a friendship, with General Leslie Groves, the U.S. Army officer who led the Manhattan Project. The two met several times a week, sometimes for a private discussion before others were allowed into the room. The general valued Chadwick's expertise, his independence from Washington politics, and his few, simple words: as Chadwick's biographer Andrew Brown puts it, "who better to keep an eye on those 'crackpots' Oppenheimer was gathering around him at Los Alamos?" Chadwick, in turn, admired Groves's drive and willingness to take risks, and was quite convinced the Manhattan Project could not have succeeded without the general's leadership.

Furthermore, Chadwick secured important concessions for the British, including a relaxation of security compartmentalization that was crucial to the free exchange of ideas and an agreement that Canada's Chalk River Laboratories could be partially integrated into the Manhattan Project, which gave Britain a head-start in reactor technology after the war. He also skillfully navigated long and tedious discussions of patents and intellectual property. When diplomacy had resolved a tricky issue, Chadwick was known to say with delight that it was "all jam and kippers" (jam and kippers, a kind of butterflied smoked herring, was considered a treat in parts of northern England).

Dawn of the Atomic Age

On July 16, 1945, Chadwick witnessed the detonation of the world's first atomic device in the New Mexico desert, known as the Trinity test. He reported later that day to London: "The test appears to have been completely successful … It was a wonderful, even fantastic experience, which I shall not attempt to describe now, as I am rather tired and should get things in wrong proportion. I am sending you this news through General Groves, who has kindly offered to convey it."

Chadwick was knighted that year by King George VI but, upon his return to the United Kingdom for the honor, he was overwhelmed by exhaustion from his work in the United States. He recovered in time to begin rebuilding the physics department at Liverpool, then moved back to Cambridge as master, or head, of Gonville and Caius College in Cambridge. He retired in 1958 and, away from the limelight he hated, he died peacefully in 1974. He was 82 years old.

James Chadwick, left, and General Leslie Groves, leader of the Manhattan Project, developed a close working relationship, often meeting for private discussions. The general valued Chadwick's expertise and diplomacy efforts.

SIR JAMES CHADWICK

1891–1974

LOS ALAMOS CONTRIBUTIONS

Leadership of the wartime British atomic energy mission to the United States.

NOBEL PRIZE: PHYSICS, 1935

Proved the existence of neutrons, which are elementary particles without electrical charge.

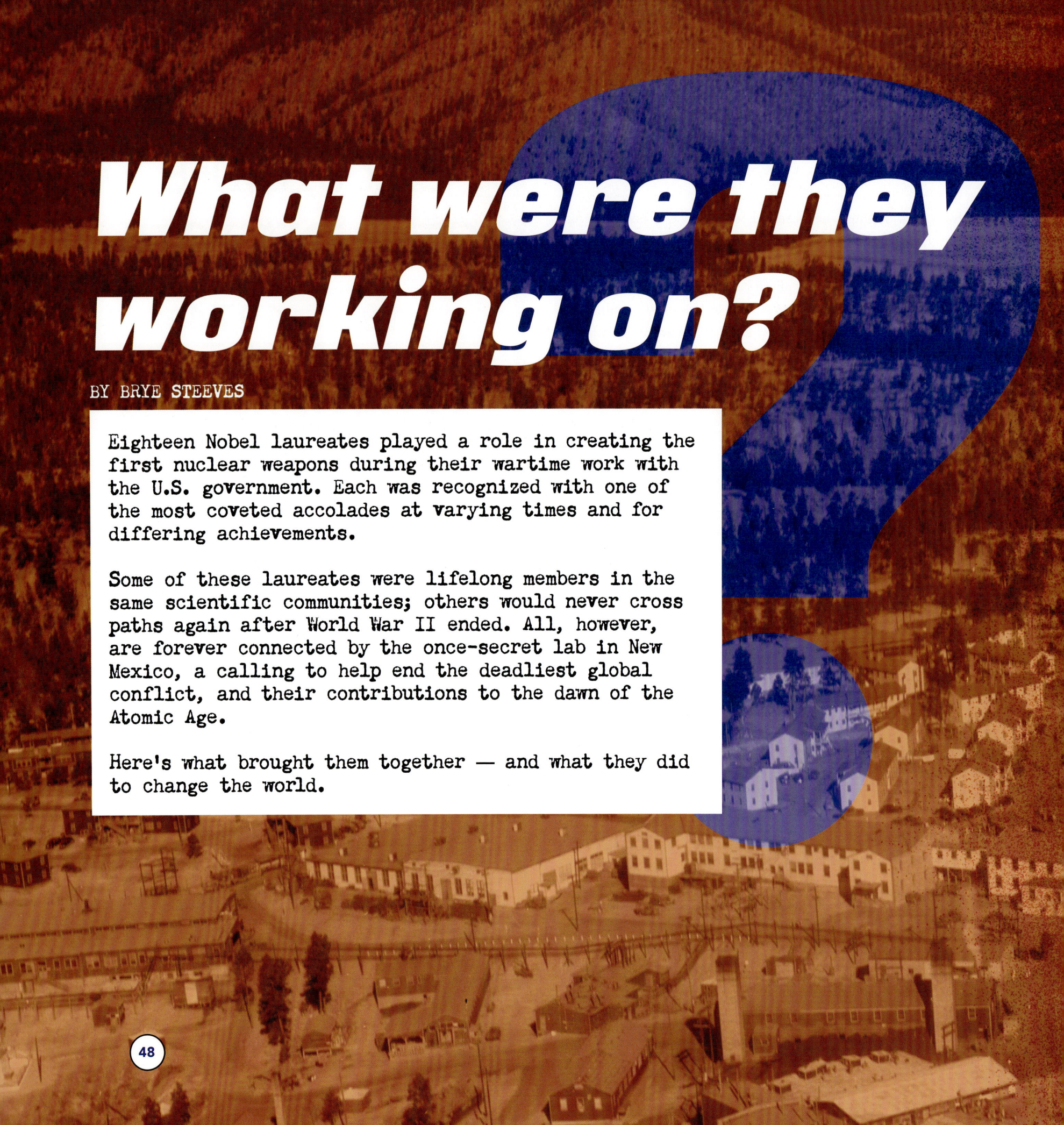

What were they working on?

BY BRYE STEEVES

Eighteen Nobel laureates played a role in creating the first nuclear weapons during their wartime work with the U.S. government. Each was recognized with one of the most coveted accolades at varying times and for differing achievements.

Some of these laureates were lifelong members in the same scientific communities; others would never cross paths again after World War II ended. All, however, are forever connected by the once-secret lab in New Mexico, a calling to help end the deadliest global conflict, and their contributions to the dawn of the Atomic Age.

Here's what brought them together — and what they did to change the world.

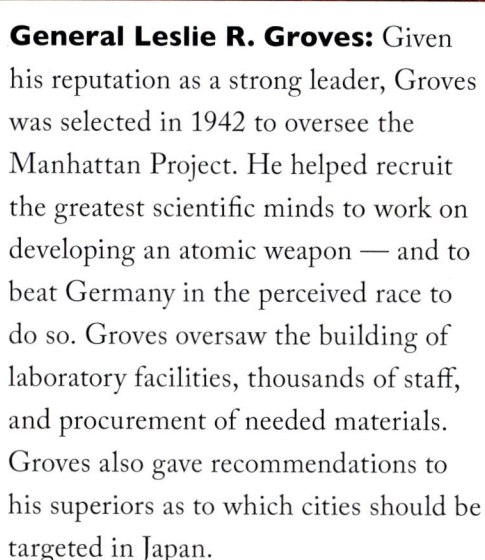

World War II: History's largest and bloodiest war formally began on September 1, 1939, with Adolf Hitler's invasion of Poland and France, and Britain's declaration of war on Germany two days later. The United States was thrust into the conflict following Japan's surprise attack on Pearl Harbor on December 7, 1941. Over the course of six years, more than 30 countries were involved and 70–85 million people died, including military personnel who were killed in combat and civilians who died as a result of wartime conditions. World War II officially ended on September 2, 1945, with Japan's surrender following the U.S. deployment of atomic bombs on the cities of Hiroshima and Nagasaki.

The Manhattan Project: The U.S. government's top-secret effort to create the world's first atomic bombs officially began on August 13, 1942, and was named for its first office in Manhattan, New York. The headquarters were moved to Washington, D.C., and project sites were later established around the country, including Los Alamos, New Mexico; Oak Ridge, Tennessee; and Hanford, Washington.

General Leslie R. Groves: Given his reputation as a strong leader, Groves was selected in 1942 to oversee the Manhattan Project. He helped recruit the greatest scientific minds to work on developing an atomic weapon — and to beat Germany in the perceived race to do so. Groves oversaw the building of laboratory facilities, thousands of staff, and procurement of needed materials. Groves also gave recommendations to his superiors as to which cities should be targeted in Japan.

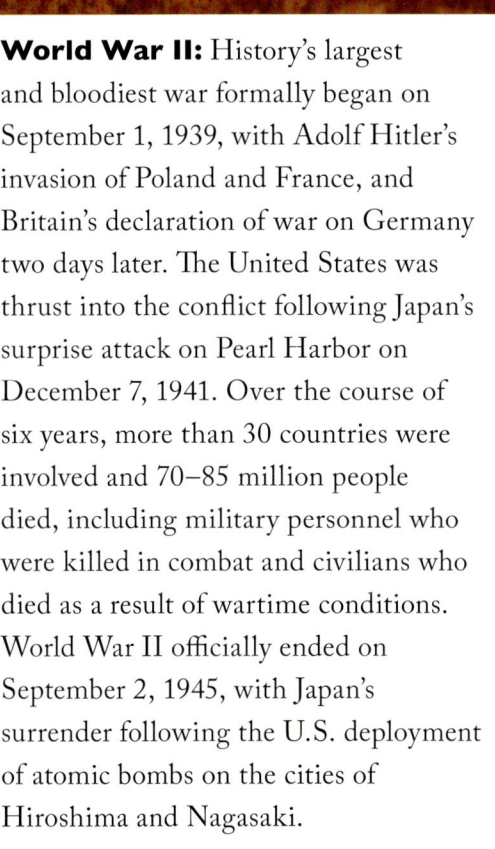

J. Robert Oppenheimer: The brilliant physicist was a professor at the University of California, Berkeley, and the California Institute of Technology, or Caltech, when Groves selected him as the Los Alamos lab director. Oppenheimer's lack of managerial experience and past communist associations did not deter Groves, who recognized Oppenheimer's scientific genius. The two men had a mutual respect for one another and a strong partnership. As the Lab's director, Oppenheimer earned a steadfast following from his subordinates who remained loyal to him when Oppenheimer's patriotism later came into question.

Los Alamos One of the earliest challenges in creating the first atomic weapons was ensuring that the laboratory location was rural enough to be shrouded in secrecy. Several areas were scouted, with Los Alamos selected based on Oppenheimer's recommendation and Grove's ultimate approval. Oppenheimer had spent time on the Pajarito Plateau as a young man — he remembered its remoteness and natural beauty. The area was relatively uninhabited, with just a small boarding school, some homesteads, and nearby Native American pueblos. Scientists, explosives experts, military personnel, support staff, and their families converged on the secret town, which grew to a population of about 7,000 by the war's end. The lab would be called Project Y and conducted both the research and construction of the atomic bombs.

The Trinity test: At around 5:30 a.m. on July 16, 1945, the world's first atomic device — nicknamed The Gadget — was detonated in southern New Mexico, proving the feasibility of weaponizing energy from the atom. The successful explosion marked the beginning of the Atomic Age. Groves later asked about the name of the test, and Oppenheimer said he believed he was inspired at the time by a line in a John Donne poem, "Batter my heart, three person'd God."

FIGURE 45   Tightening the Polar Cap to the Zone Segments.

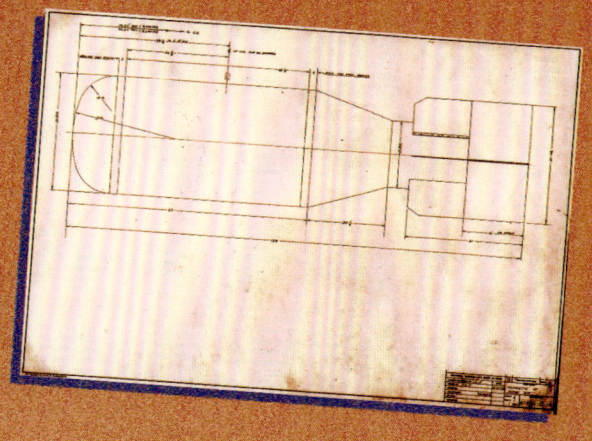

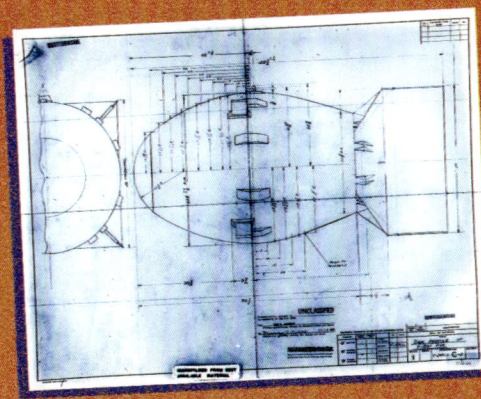

The Gadget: The world's first-ever atomic device was detonated in the New Mexico desert during the Trinity test. It verified that an implosion-type plutonium bomb (Fat Man) would be successful when released above Nagasaki just weeks later. The power from The Gadget's detonation was equivalent to around 21,000 tons of TNT; its mushroom cloud grew to about 3,280 feet wide with a column of smoke at a height in excess of 40,000 feet.

Little Boy: The first of two atomic bombs to be used in combat, the uranium gun-type weapon, was released above Hiroshima on August 6, 1945. Little Boy was developed after scientists realized "Thin Man" (a plutonium gun-type weapon) would not be successful. Thin Man is thought to have been named in a nod to President Franklin Roosevelt, and Little Boy was likely coined as such because it was smaller in size, though the weapon was 9,700 pounds, 10 feet long, and just over 2 feet in diameter.

Fat Man: The second of two atomic bombs to be used in combat, the plutonium implosion-type weapon was released above Nagasaki on August 9, 1945. The bulbous-looking Fat Man is said to have been named after British Prime Minister Winston Churchill. The bomb was a weaponized version of The Gadget that was detonated during the Trinity test in the New Mexico desert. Fat Man was 10,800 pounds, nearly 11 feet long, and 5 feet in diameter.

After the war to today: The scientific achievements from the Los Alamos wartime lab were remarkable and form today's legacy. Following World War II and a shifting national security mission, the no-longer-secret lab has continued to lead advancements in weapons technology during the past eight decades. A safe, reliable, and effective deterrent is based on innovative science and cutting-edge technology that began in 1943 and will endure into the future.

OWEN
CHAMBERLAIN

By Roger Meade

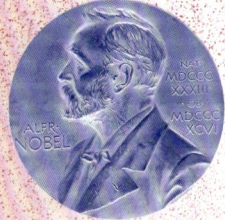

"By Jove, this is what I want to do." And so began Owen Chamberlain's successful quest to verify the existence of the antiproton, a proton having a negative electrical charge. First proposed by theoretical physicist Paul Dirac in 1930, the antiproton remained an elusive construct until Chamberlain and Emilio Segrè proved its existence in 1955. Their discovery, published in *Physics Review*, was followed by the 1959 Nobel Prize in Physics.

Born in 1920 in San Francisco, Chamberlain showed an early aptitude for both mathematics and physics, graduating with honors in 1941 from Dartmouth College in Hanover, New Hampshire. With Dartmouth's Thayer Prize for Mathematics in hand, Chamberlain began graduate work at the Berkeley campus of the University of California. When World War II broke out, he abandoned his studies, joining a research team investigating spontaneous fission in uranium. This team, led by Segrè (see Page 136), moved to Los Alamos in early 1943, where Chamberlain contributed to the discovery of a much higher than predicted spontaneous fission rate in plutonium.

This finding caused a major reorganization of Los Alamos in August 1944, leading to the development of Fat Man, which was the bomb type tested at the Trinity site in the New Mexico desert in July 1945. The test was the world's first successful detonation of a nuclear weapon and marked the dawn of the Atomic Age. The same plutonium implosion-style bomb would be released just weeks later above Japan as one of two Los Alamos-created weapons to help bring a swift end to World War II.

After Los Alamos
After the war ended in September 1945, Chamberlain enrolled at the University of Chicago, where he studied under Nobel laureate and fellow Los Alamos scientist Enrico Fermi (see Page 64).

The Fat Man atomic bomb is secured onto a trailer cradle for delivery and loading onto the aircraft that would carry it to its target. It was released above Nagasaki, Japan on August 9, 1945. It was the second of two nuclear weapons to ever be used in combat.

In 1948, while still working on his doctorate, Chamberlain accepted an appointment as an instructor at Berkeley, where he would spend the rest of his professional career.

Unpretentious and approachable, he was known as just "Owen" and was a popular professor with his undergraduate and graduate physics students at Berkeley, according to a written tribute published by the university after his death. Colleagues recalled that during Chamberlain's office hours, his space often couldn't accommodate all the students who came to see him.

Chamberlain's early research at Berkeley focused on proton-proton scattering, work that contributed to his hypothesis that the mass and charge of an antiproton could be measured by its momentum and velocity. When well-known physicist Maurice Goldhaber made a public bet that the antiproton did not exist, Chamberlain decided then and there to prove his colleague wrong.

In 1941, Owen Chamberlain began graduate school at the University of California, Berkeley, and joined the Manhattan Project the following year.

The seminal experiment that led to the discovery of the antiproton began in the summer of 1955. As later described by a colleague, Herbert Steiner:

> The proton beam in the Bevatron was used to produce secondary particles in a copper target. A series of magnets then transported a negatively charged beam of known momentum to the velocity-defining detectors. Two scintillation counters, separated by 13 meters, were used to measure the 13 nanosecond time difference between the rare antiprotons and the much more copiously produced pions in the beam.

Chamberlain later said,

> Fortunately the antiprotons were slightly more numerous [than predicted], "being 1 in 30,000 particles [rather than the predicted 1 in 100,000] in the magnetized beam. If there had been appreciably fewer antiprotons, we might altogether have missed seeing them on the first try. As it was, we saw only one antiproton for every fifteen minutes with the first apparatus.

By experimentally proving Dirac's theory, the discovery of the antiproton advanced the fundamental understanding of nature and its constituent particles and led directly to the discovery of a third subatomic particle, the antineutron.

And, Maurice Goldhaber paid off his bet.

Later in life

After the discovery of the antiproton, Chamberlain devoted much of his remaining career to polarization studies, including the development of polarized targets that became a standard part of particle physics research worldwide. Chamberlain then conducted the first use of these new targets for experiments "using high energy proton and pion beams." Among the last of his intensive

Owen Chamberlain came to Los Alamos from the University of California and worked on the Fat Man atomic bomb.

Chamberlain, Owen & wife   From: Berkeley
Married: yes
Arrival: 5/31-6/6   Salary: $260.00
          6/29 to Frijoles

Dr. W. Edward Chamberlain, 3301 N. Broad St., Phila.
Mrs. G. H. Copper, 2280 Allesandro St., Los Angeles

    July 8 to Site

wife Babette C. Chamberlain
    with Segre   Terminated: Lst. wk. day3-12-46
                        Terminated 3-20-46

When well-known physicist Maurice Goldhaber made a public bet that the antiproton did not exist, Chamberlain decided then and there to prove his colleague wrong.

Owen Chamberlain sits behind the wheel of a jeep, a common way to traverse the many unpaved roads in Los Alamos during the Manhattan Project era. The laboratory was quickly constructed as scientists, engineers, military personnel, support staff, and their families moved to the the area in support of the atomic bombs' creation.

research activities was work at the Stanford Linear Accelerator Center to improve the precision of "the Standard Model of electro weak interactions."

Although advancing age and Parkinson's disease slowed down Chamberlain's physical activities, he remained intellectually active up to his death on February 28, 2006. He was 85 and had attended a departmental physics colloquium at Berkeley just days earlier. Chamberlain was preceded in death by his first wife in 1988 and his second wife in 1991. He was survived by his third wife, Senta, and four children, Karen, Lynne, Pia, and Darol, from his first marriage.

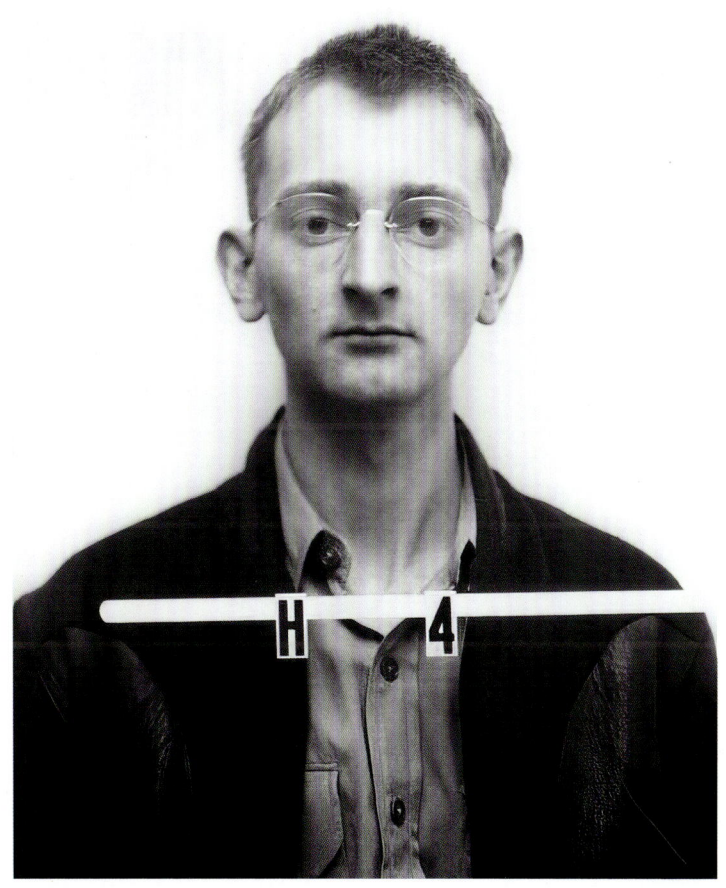

Owen Chamberlain's badge photo was taken shortly after his arrival to northern New Mexico to work on the creation of the atomic bombs at the top-secret lab in Los Alamos.

OWEN CHAMBERLAIN

1920–2006

LOS ALAMOS CONTRIBUTIONS

Contributed to the discovery of a much higher than predicted spontaneous fission rate in plutonium. This finding caused a major reorganization of Los Alamos in August 1944, leading to the development of the Fat Man-type nuclear weapon.

NOBEL PRIZE: PHYSICS, 1959

Jointly awarded to Chamberlain and physicist Emilio Segrè for verifying the existence of the antiproton, which is a proton having a negative electrical charge.

ENRICO FERMI

By Madeline Whitacre
and Amy Belotti

He wasn't Jewish, but Enrico Fermi needed to flee Mussolini's Fascist Italy. It would be his scientific achievements that would allow him to do so — being awarded the Nobel Prize in Physics ultimately saved his family's life. In late 1938, Fermi left Italy to collect his prize at the annual ceremony in Stockholm, Sweden, taking with him his wife Laura, who came from a prominent Jewish family, and their two young children, Nella and Giulio.

From there, the family quietly moved to the United States, escaping persecution and finding freedom in New York, where Fermi was hired as a physics professor at Columbia University.

Not long after, Fermi's work led him to a top-secret lab in the mountains of northern New Mexico, where his scientific genius would ultimately play a role helping end the bloodshed of World War II.

Early years

Fermi was born on September 29, 1901 in Rome. His aptitude for math and physics was apparent early on. When he was 17 years old, he won a fellowship to study at the University of Pisa. When he was only 20, he graduated with a degree in physics and then accepted a position working with the famous German physicist Max Born in Göttingen for a semester.

When he returned to Italy from Göttingen, Fermi met 16 year-old Laura Capon at a soccer game with friends. He again left Italy and moved to Leyden to work with Paul Ehrenfest for a semester, but didn't forget her. Fermi went back to Italy and, in 1926, became a professor of theoretical physics at the University of Rome around the same time Laura Capon was starting school there. In 1928, they were married.

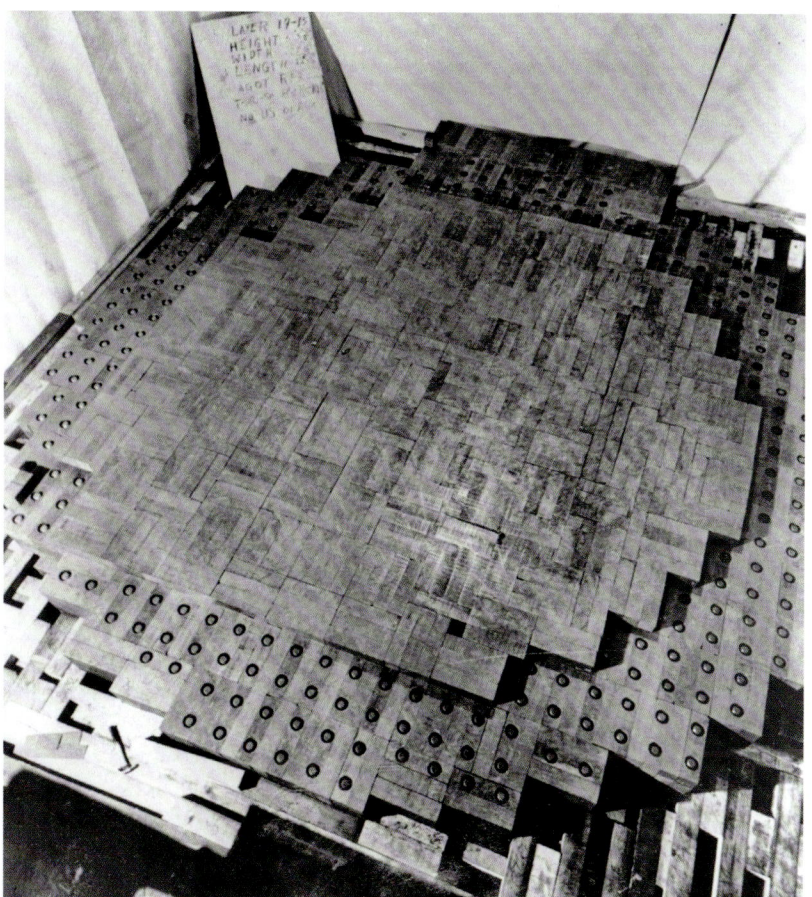

Chicago Pile-1, shown here, produced the world's first manmade controlled nuclear chain reaction. Enrico Fermi led the experiment on December 2, 1942. Its success was the precursor to the development of the first atomic weapons.

Scientific contributions

Eventually, Fermi began to focus his research on the atomic nucleus, contributing to beta decay theory. In 1934 Irène and Frédéric Joliot-Curie discovered artificial radioactivity, and Fermi closely followed these developments. Following this discovery, Fermi began conducting his own experiments in which he bombarded various elements with neutrons. This made many of the elements radioactive and Fermi noticed that sometimes the level of radioactivity increased greatly — for example, when the neutrons first passed through paraffin the material slowed down the neutrons. Fermi's discovery of slow neutrons would help pave the way for future developments in atomic physics.

In 1938, Fermi was awarded the Nobel Prize "for his demonstrations of the existence of new radioactive elements produced by neutron irradiation, and for his related discovery of nuclear reactions brought about by slow neutrons."

Following their escape from Fascist Italy, the Fermi family settled in New York where Enrico Fermi was hired at Columbia University. There, Fermi worked on fission research and began investigating the possibilities of a chain reaction using uranium.

Chicago Pile-1

Near the end of 1941, Arthur Compton, a member of the Office of Scientific Research and Development's Advisory Committee on Uranium, relocated the reactor work from Columbia to the University of Chicago. Fermi moved to Chicago to join the university's Metallurgical Laboratory. There, he began work on a graphite "pile," an experiment that would use graphite as a moderator for the uranium. Moderators are substances like graphite and paraffin that slow down neutrons, which lose energy in collisions among atoms in the moderator. Construction for Chicago Pile-1, or CP-1, began in November 1942. The pile was built in the squash courts underneath the university's Stagg Field. On December 2, Fermi led the experiment and CP-1 went critical, meaning it was the first manmade controlled nuclear chain reaction.

The next step to develop an atomic bomb was to create an uncontrolled nuclear chain reaction, which means the reaction would produce a large amount of energy all at once. This led to the establishment of the Los Alamos laboratory as part of the U.S. government's Manhattan Project. Many of the scientists who were a part of the CP-1 experiment were recruited to the top-secret lab in northern New Mexico, where they would work to create the first nuclear weapons.

While at the laboratory, Enrico Fermi created a handheld analog computer: the FERMIAC. This tool demonstrated how neutron behavior could be modeled.

The lab's first director, J. Robert Oppenheimer, recruited Fermi as a division leader and an associate director. Named for Fermi, the lab's new F Division was established in September 1944.

Two of the groups in F Division conducted early work on the hydrogen bomb, or Super. F-1 was the Super and General Theory Group led by Edward Teller and F-3 was the Super Experimentation Group led by Egon Bretscher. This work would become an important foundation for thermonuclear design work after World War II. Fermi also managed the Water Boiler Group and the Fission Studies Group.

In addition to his position as a division leader, Fermi conducted theoretical and experimental research. He worked on the calibration of neutron sources, cross-section measurements, and alpha absorption. One particular experiment highlighted Fermi's ingenuity. On the morning of July 16, 1945 Fermi was at the Trinity test site and observed the world's first nuclear detonation:

> About 40 seconds after the explosion the air blast reached me. I tried to estimate its strength by dropping from about six feet small pieces of paper before, during and after the passage of the blast wave. Since, at that time, there was no wind I could observe very distinctly and actually measure the displacement of the pieces of paper that were in the process of falling while the blast was passing. The shift was about 2 ½ meters, which, at the time, I estimated to correspond to the blast that would be produced by ten thousand tons of T.N.T.

In actuality, the Trinity test produced a yield equivalent to 21 thousand tons of T.N.T. Fermi's estimate, off by only about a factor of two, was still remarkably close given that his only tools of measurement included a few scraps of paper.

Post war

After the end of World War II, Fermi left the laboratory in December 1945, although F Division had been dissolved in October. The Fermi family returned to Chicago, where Fermi became a professor at the University of Chicago's newly-established

Enrico Fermi won the Nobel Prize in 1938 for his identification of new radioactive elements. He left Italy to collect his prize and quietly fled to the United States. (Photo courtesy of the Department of Energy.)

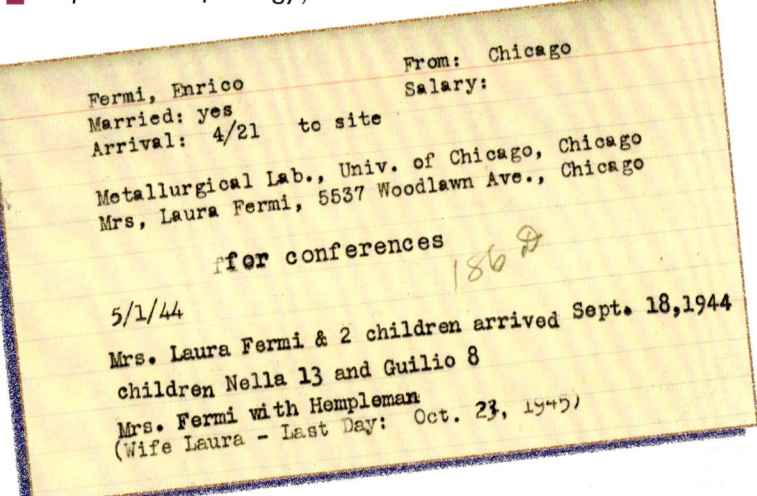

Enrico Fermi's lab employment card documented his and his family's arrival to Los Alamos in September 1944, and Fermi's earlier visit for the Los Alamos premier conference.

Institute for Nuclear Studies. He worked both at the university and at nearby Argonne National Laboratory, the successor to the Metallurgical Laboratory. Fermi served on the first General Advisory Committee (GAC) to the newly-formed Atomic Energy Commission that would manage atomic energy research and development in the United States. In this position, Fermi and his fellow committee members opposed a crash program to develop a hydrogen bomb, though President Truman advanced the effort. Fermi served one term on the GAC from January 1, 1947 to August 1, 1950.

Around this same time, Fermi returned to Europe for the first time since fleeing with his family more than a decade earlier. He attended various lectures and conferences, including a conference on cosmic rays, an area in which he had published papers.

Fermi led the experiment and CP-1 went critical on December 2, making it the first manmade controlled nuclear chain reaction.

Enrico Fermi, second from the left, often found time to hike with colleagues in the area surrounding the secret Los Alamos laboratory. He also enjoyed outdoor winter activities, like skiing. He lived in Los Alamos with his wife and their daughter and son. (LANL photo by L.D.P King.)

Fermi continued experimenting, conducting research, and lecturing until an unknown illness prompted an exploratory surgery in the fall of 1954. Fermi died from stomach cancer on November 29, 1954 at the age of 53.

Enrico Fermi's Los Alamos badge photo was taken upon arrival in New Mexico. Fermi worked at the Los Alamos Lab until shortly after World War II ended.

ENRICO FERMI

1901–1954

LOS ALAMOS CONTRIBUTIONS

Served as a division leader and an associate director. Named for Fermi, the lab's F Division included work on the hydrogen bomb.

NOBEL PRIZE: PHYSICS, 1938

For demonstration of the existence of new radioactive elements produced by neutron irradiation and his related discovery of nuclear reactions associated with slow neutrons.

RICHARD FEYNMAN

By John Moore

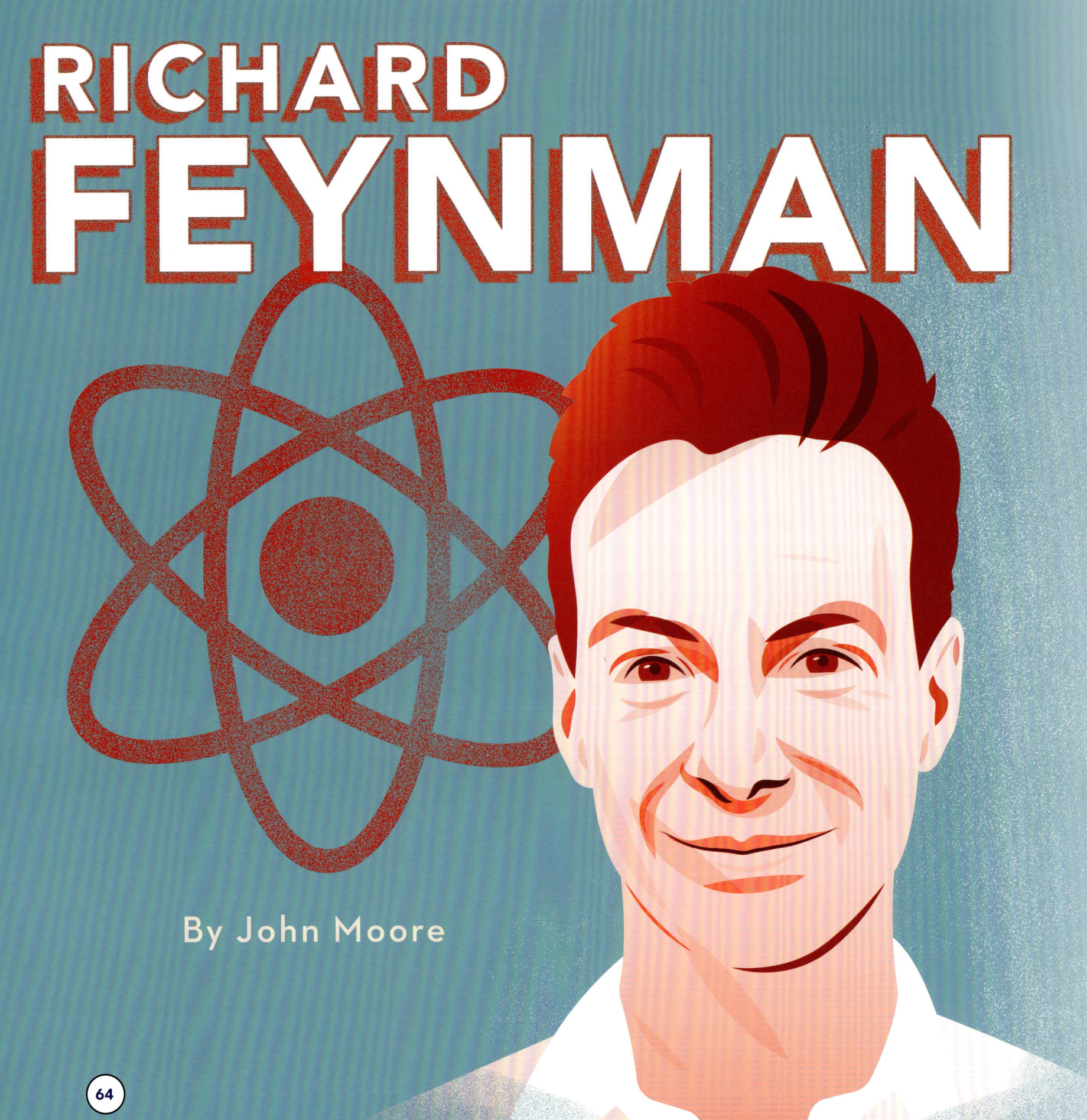

Shortly after his death in February 1988, *The New York Times* proclaimed him as, "arguably the most brilliant, iconoclastic, and influential of the post-war generation of theoretical physicists." Many in Los Alamos during the lab's first years would remember his quick wit, humor, and practical jokes, including cracking combinations on scientists' top-secret safes and pranking the lab's security guards.

Perhaps it's being the genius jokester that defines the legacy of Richard Feynman, not to mention the 1965 Nobel Prize in Physics that was followed by an acceptance speech filled with quips and marked with laughs.

His work ranged from atomic weapons development to NASA consultation to Nobel Prize-winning contributions in quantum electrodynamics, but his personality also made its mark. Physicist Philip Morrison described Feynman as "the most original theoretical physicist of our time" who also "liked colorful language and jokes."

Early years

Born in Queens, New York, on May 11, 1918, Feynman had a curiosity in science at an early age. This innovative mindset was evident when he set up a laboratory in his childhood home to conduct experiments. His curiosity continued into his early teens when he became interested in repairing and taking apart radio sets. He said, "Usually they were broken in some simple-minded way, some obvious wire was hanging loose, or a coil was broken or partly unwound, so I could get some of them going." Feynman's aptitude for mathematics was also apparent — in high school, he taught himself trigonometry and calculus.

After graduation, Feynman attended the Massachusetts Institution of Technology (MIT) in Cambridge to study mathematics and electrical engineering. He received his bachelor's degree in 1939 and went on to earn a Ph.D. in quantum mechanics in 1942 at Princeton, with

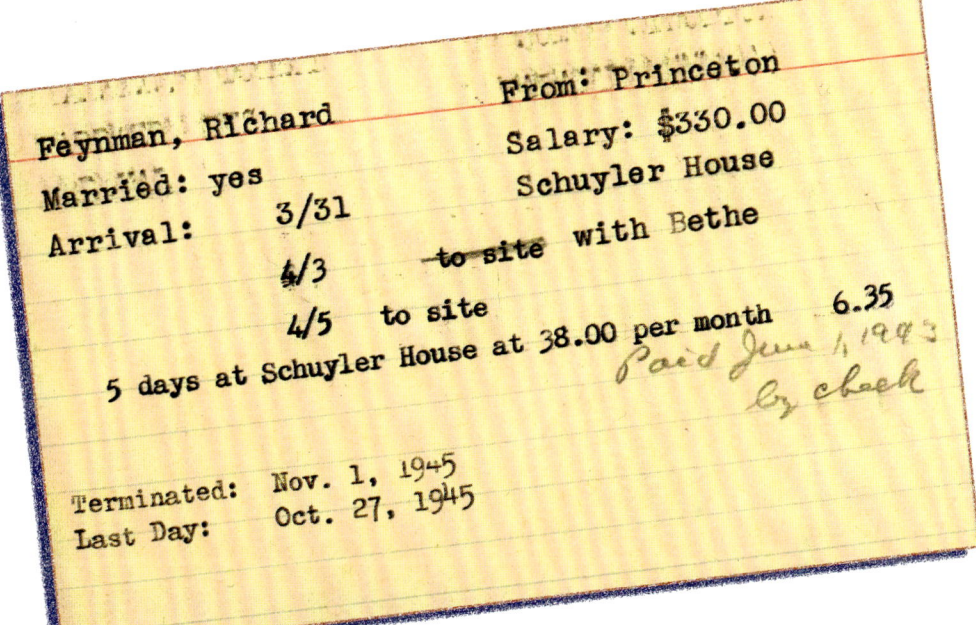

Incoming Manhattan Project staff were issued these 4-inch by 5-inch index cards, which included personal information, as part of their inprocessing. Feynman moved to Los Alamos with his wife Arline, whose health was deteriorating due to tuberculosis. Sadly, she died in an Albuquerque hospital not long after their arrival.

professor John Wheeler. While working on his Ph.D., Feynman married Arline Greenbaum who was suffering from tuberculosis.

Los Alamos work

With the United States' entrance into World War II, after Japan's surprise attack on Pearl Harbor on December 7, 1941, Feynman began working with physicist Robert Wilson at Princeton University in New Jersey on the separation of uranium. In 1943, as a part of Wilson's team, Feynman arrived at the then-secret lab in Los Alamos to take part in the Manhattan Project, which was the United States-led effort to create the world's first atomic weapons.

While at Los Alamos, Feynman worked with physicist Hans Bethe, who described Feynman as the most ingenious member of his team, according to *The New York Times*. In the Theoretical Division, staff researched and calculated the efficiency of the nuclear devices. Feynman monitored and conducted experiments on a nuclear reactor known as the Water Boiler, which provided controlled nuclear reactions for research.

Starting in the spring of 1944, Feynman played an important consulting role in the assembly and use of IBM electromechanical punch-card accounting

The Jumbo device was designed to prevent the loss of plutonium if the Trinity test was not successful. By test time, scientists, including Richard Feynman, were confident and it wasn't used.

equipment at the lab. The machines performed vital implosion hydrodynamics calculations needed for atomic-bomb design selection and development.

Collaborating with the equipment operators, Feynman helped to refine the operating process on the machines, reducing the time needed to complete an implosion calculation from three months to only three weeks.

While at Los Alamos, his wife's health continued to deteriorate from tuberculosis and she was admitted to a hospital over 90 miles away in Albuquerque. As often as possible, Feynman visited Arline — borrowing a car from his lab colleague Klaus Fuchs, who would later be confirmed as a Soviet spy.

Sadly, Arline died June 16, 1945. Feynman was in Albuquerque with her during her final hours. Years later, Feynman wrote her a letter that was discovered by a biographer among Feynman's documents after his death in 1988. Feynman wrote, "I find it hard to understand in my mind what it means

to love you after you are dead — but I still want to comfort and take care of you — and I want you to love me and care for me. I want to have problems to discuss with you — I want to do little projects with you."

After Arline's death, Feynman continued his work at the Los Alamos lab. Exactly one month later, on July 16, 1945, Feynman witnessed the Trinity test, the world's first detonation of an atomic weapon, alongside his lab colleagues in the New Mexico desert. Although observers were given safety glasses, Feynman chose not to wear his and afterward stated, "I am about the only guy who actually looked at the damn thing."

After World War II and a Nobel Prize

When the war ended, Feynman left Los Alamos in October 1945 to teach physics at Cornell University in Ithaca, New York, for several years. This position was followed by a longer stint at the California Institute of Technology (Caltech).

In 1953, Feynman published several papers on helium-3 (an isotope of helium) in which he explained how the atoms moved and why the curve that describes the energy of their excitations had a peculiar and unusual shape. No one previously had been able to show why this curve had its particular shape. Feynman's work provided a foundation of how superfluidity, or flowing without friction, works. Although groundbreaking, it would be other work for which Feynman would later be recognized with the world's top prize.

The first page of Richard Feynman's personnel paperwork lists the now-famous address P.O. Box 1663 Santa Fe, New Mexico, as his residence. The address was used for all official documents as well as mail in Los Alamos in an effort to keep the work there secret.

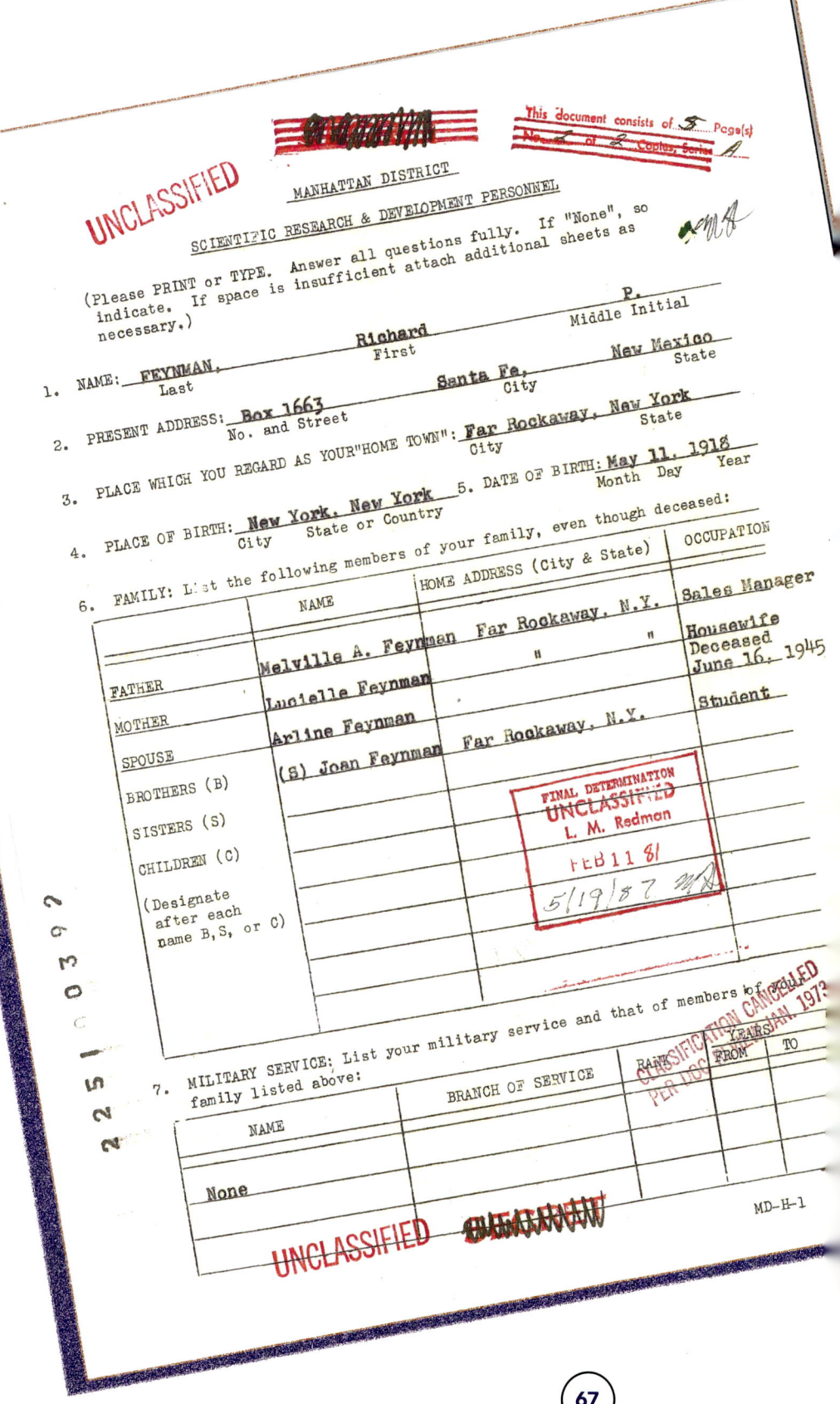

"Nobody ever figures out what life is all about, and it doesn't matter," Feynman once said. "Explore the world. Nearly everything is really interesting if you go into it deeply enough."

Meanwhile, after a short second marriage, Feynman married a third time to Gweneth Howarth in 1960. The couple had a son, Carl, and adopted a daughter, Michelle.

In 1965, Feynman, Sin-Itiro Tomonaga of Japan, and Julian Schwinger of Harvard University were jointly awarded the Nobel Prize in Physics for their work in quantum electrodynamics. Quantum electrodynamics is the study of physics that illustrates the interaction of matter and light. It explains occurrences between charged particles through the exchange of protons and the shifts in energy and particles in physics.

After receiving his Nobel Prize, Feynman used the monetary portion of the award to purchase a home in Baja California, Mexico. Feynman would also develop many pastimes, which included an interest with isolation tanks, also known as float therapy. During one of Feynman's trips to Brazil, he became fascinated with the traditional instrument called the frigideira and Samba music. He joined a Samba music school and learned the instrument, which looks like a frying pan and is hit with a stick, and eventually played in the school's band for Rio de Janeiro's annual Carnival festivities.

By the late 1970s, Feynman was seeking treatment for a rare cancer that develops in fatty tissue known as liposarcoma. He was, however, still working on significant projects, including his 1985 book of playful reminiscences *Surely You're Joking, Mr. Feynman*. It was on *The New York Times* best-seller list for 14 weeks.

In 1986, Feynman served on the Rogers Commission, which investigated the cause of the Space Shuttle Challenger explosion shortly after liftoff. He caused a stir during the televised hearing when he dropped an O ring in a cup of ice water to demonstrate the detrimental effect of a cold temperature on the effectiveness of the rubber material.

Not long after, in the winter of 1988, Feynman died from kidney failure. He was 69 years old and survived by his wife and children.

Richard Feynman (center) talks with mathematicians Stan Ulam (right) and John von Neumann (left). The Manhattan Project's lab in Los Alamos brought together the world's greatest minds, working toward one goal: to create the world's first nuclear weapons in a perceived race against Nazi Germany to do so. After just 27 months, the United States released two atomic bombs above Japan, helping to end World War II shortly thereafter.

Richard Feynman along with scientist Charles Critchfield and lab Director J. Robert Oppenhiemer attend a post-war party in 1945. Feynman, as did the majority of the scientists, left Los Alamos not long after the official end of World War II on September 2.

RICHARD FEYNMAN

1918–1988

LOS ALAMOS CONTRIBUTIONS

Worked with Hans Bethe in the Theoretical Division (T Division). Conducted and monitored the Water Boiler reactor.

NOBEL PRIZE: PHYSICS, 1965

Feynman along with Sin-Itiro Tomonaga and Julian Schwinger were awarded the prize for their work in quantum electrodynamics (QED).

Nobel Laureates* of the Manhattan Project's lab in
LOS ALAMOS

Niels Bohr
1922

Sir James Chadwick
1935

Enrico Fermi
1938

Isidor Rabi
1944

Edwin McMillan
1951

Felix Bloch
1952

Emilio Segrè
1959

Owen Chamberlain
1959

Maria Goeppert Mayer
1963

1920 **1930** **1940** **1950** **1960**

* Each laureate appears with the category of their prize and year they received it.

1939
World War II starts

1941
United States enters into World War II

1943
Los Alamos Manhattan Project begins

1945
World War II ends

1946
Manhattan Project ends

Richard Feynman 1965

Hans Bethe 1967

Luis Alvarez 1968

Aage Bohr 1975

Val Fitch 1980

Norman Ramsey 1989

Joseph Rotblat 1995

Frederick Reines 1995

Roy Glauber 2005

1970　　**1980**　　**1990**　　**2000**　　**2010**

VAL FITCH

By Jacqueline Kilby

It is highly improbable, Val Fitch said, to begin life on a cattle ranch in Nebraska and then nearly 60 years later be in Sweden to accept the Nobel Prize in Physics. Fitch did just that. However, his course of events indeed became probable, he said, thanks to the people in his life. Those people included family and friends he met at the Los Alamos laboratory, including his first wife, Elise Cunningham, and the best man at their wedding, Hans Courant.

Fitch also met important colleagues during the Manhattan Project era, such as Robert Bacher, the leader of the Physics Division at the lab, who influenced Fitch's career decision to further study electrical engineering and then move on to experimental physics.

Fitch was sent to work at the then-secret lab as a member of the Army's Special Engineer District (SED). He later described this work with the Manhattan Project, which was the U.S. government's top-secret effort to create the world's first nuclear weapons, as "highly stimulating" and "probably the most significant occurrence in my education." After World War II ended and when the Manhattan Project concluded shortly thereafter, Fitch was inspired to return to school in pursuit of electrical engineering and translate that knowledge into furthering high-energy physics.

Later, his studies alongside fellow physicist James Cronin ushered in what many consider the golden age of particle physics by using a cosmotron particle accelerator machine to measure the lifetimes of K-mesons. Fitch's obsession with K-mesons, which are subatomic particles with a mass between an electron and a proton that bind nucleons together in the atomic nucleus, lasted his lifetime and earned Fitch and Cronin the 1980 Nobel Prize in Physics. They discovered that two quantities differed by a few percent, indicating a small and direct symmetry violation showing that the left-right asymmetry is not always completely compensated by transforming

Trinitite, shown enlarged here, is a glassy residue that remained on the New Mexico desert floor after Trinity, which was the first-ever detonation of an atomic bomb. Trinitite is radioactive, but safe to touch.

from matter to antimatter. These experiments led directly to major progress in elementary particle physics and theories on how the matter of the universe survived the Big Bang.

Early years

Fitch was born on a cattle ranch in Merriman, Nebraska, near the site of the Battle of Wounded Knee, in which nearly 300 of Lakota or Sioux were killed by the U.S. Army in 1890. His parents purchased the property just two decades after the massacre, and the tribe was a large part of his early life. Fitch built a laboratory in the basement of his family home that consisted of a chemistry set, a radio, and other items that helped him learn about the world. After graduating as valedictorian from Gordon High School and attending two and a half years at Chadron State College in Nebraska, Fitch was drafted into the Army in 1943; World War II had already been underway for several years.

Los Alamos work

Fitch completed basic military training in Kearns, Utah, and was sent to the Army Specialized Training Program (ASTP) in

Pittsburgh at Carnegie Institute of Technology, which later became Carnegie Mellon University. The program was designated for those with technical skills.

Fitch concentrated on chemistry, physics, and mathematics in his undergraduate work at Chadron State College, so he chose electrical engineering as his focus in ASTP. The pressure for more manpower in the battlefield mounted, and by 1944 the ASTP units were disbanded. Most of Fitch's friends from the program went off to the European theater in the 95th Infantry. The few left behind from the former ASTP were sent to Los Alamos by the end of 1944. Fitch joined the SED and was assigned to the British Mission under nuclear physicist Ernest Titterton.

"Ernie's team," as it was known, was responsible for developing the timing apparatus for the nuclear weapons implosion program, as well as the electronics for the measurement of the spherical shockwave that passes particular landmarks and physical points after the blast.

In the spring of 1945, Fitch was sent to Wendover Field near the Nevada-Utah border to work on the dummy bomb tests dropped on targets in the Salton Sea lake in California. Titterton's team was installing an apparatus to trigger the detonation and record the data from the testing, specifically measuring the simultaneous firing of detonators. Shortly thereafter, Fitch was sent to Alamogordo, New Mexico, to work on the Trinity test. The lab's staff detonated the world's first nuclear weapon at the site on July 16, 1945, marking the beginning of the Atomic Age and helping to end World War II weeks later. After witnessing the historic event that day in the desert, Fitch remarked that the filler used to insulate the recording-apparatus bunker was completely transparent, with some of the sand turning to glass from the heat of the explosion. The recording apparatus bunker was protected by transparent lead glass and the heat from the explosion first melted the sand and formed the glassy mineral trinitite, which was what Fitch was seeing.

Fitch, Val Logsdon
Arrival: May 7, 1948

With: John K. Lamb
Terminated: Last day of work August 27, 1948
Fwd. Address: Gordon, Nebraska

Val Fitch was an enlisted soldier in the Special Engineer Detachment (SED) at the Los Alamos lab during the Manhattan Project, which was the United States-led effort to create the first-ever atomic bombs and help end World War II. Fitch was discharged in 1946, and he worked as a civilian at the laboratory for another year.

IM-9:C76

Val Fitch and his colleagues witnessed the Trinity test on July 16, 1945, from Sandia Peak, approximately 150 miles away from the test site in the northern New Mexico desert. (LANL photo by Jack Abey edited by Peter Kuran.)

During his time at the lab, Fitch met Elise Cunningham. She was working as a secretary and dating his friend, Hans Courant, who was a fellow SED and a German-born American physicist. Courant left Los Alamos soon after the end of World War II to finish his college work at Massachusetts Institute of Technology (MIT). Fitch claims Courant "assigned me the job of taking care of his girlfriend … This I did, and we eventually were married … that was a nice story." After a few years of dating, Val and Elise married in 1949. Courant attended the wedding as Fitch's best man. The couple had two sons and were married until Elise died in 1972. He later married Daisy Harper in 1976.

Post-World War II

The Army discharged Fitch in 1946 and he worked as a civilian at the laboratory for another year. Working on the Manhattan Project inspired Fitch, and he went on to finish his undergraduate degree in electrical engineering from McGill University in Montreal, Canada. He then accepted a position at Columbia University in New York City with Manhattan Project physicist and future Nobel laureate Leo Rainwater. By 1954 Fitch had obtained his Ph.D. in physics from Columbia University, and was then recruited by Princeton University, where he focused on experimental and high-energy physics. A 1963 experiment that Fitch and his co-researcher James Cronin conducted revealed that matter and antimatter obeyed slightly different laws of physics. Fitch and Cronin discovered that energy symmetry was not conserved in weak interactions with two K-meson-like charged particles of the same mass. It was this work that earned Fitch and Cronin the Nobel Prize in 1980.

Around this time, from 1976 to 1981, Fitch served as chairman of the physics department at Princeton University where he earned multiple prestigious accolades.

He retired from Princeton in 1993, but continued writing articles on particle physics and contributing to the physics community. Fitch died on February 5, 2015, at the age of 91. He was survived by those he credited with his not-so-probable ascension from a cattle ranch in Nebraska to a Nobel Prize winning physicist, as well as his second wife, Daisy, and one of his sons, Alan.

Physicists Leo Rainwater and Val Fitch inspect a piece of equipment in the Columbia University laboratory. Rainwater was awarded the Nobel Prize in Physics in 1975, with Fitch following him five years later. (Columbia University, courtesy of AIP Emilio Segrè Visual Archives.)

VAL FITCH

1923–2015

LOS ALAMOS CONTRIBUTIONS

Member of the Special Engineer Detachment (SED). Fitch worked on the apparatuses to trigger the detonation of and record the data from testing.

NOBEL PRIZE: PHYSICS, 1980

Awarded "for the discovery of violations of fundamental symmetry principles in the decay of neutral K-mesons." Val Fitch and James Cronin discovered that matter-antimatter symmetry is violated when the neutral K-meson decays.

ROY GLAUBER

By Jacqueline Kilby

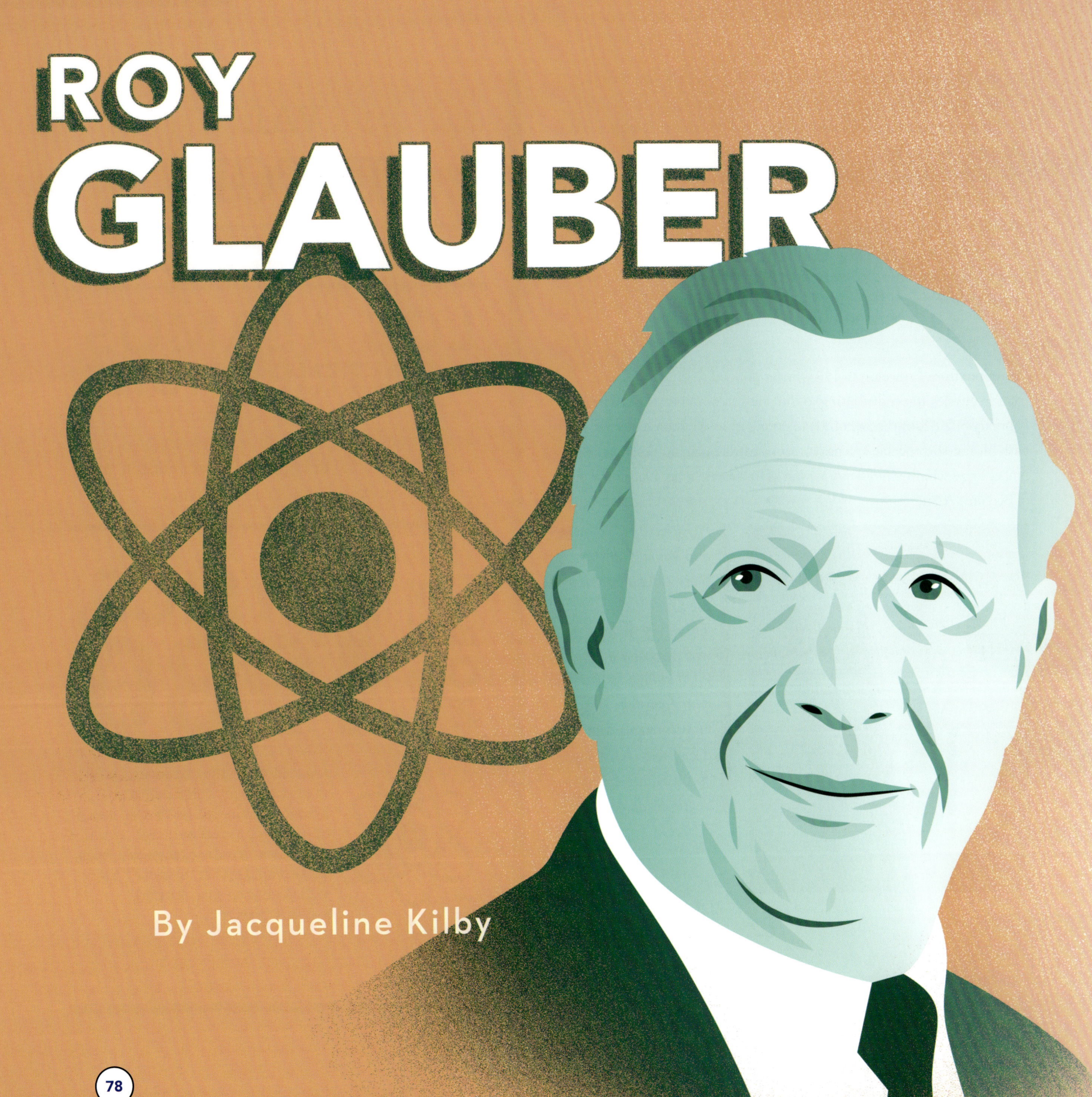

At 18 years old, Roy Glauber was one of the youngest scientists tapped to join the U.S. government's top-secret Manhattan Project. It was the fall of 1943 and without a college degree (he was just a sophomore at Harvard University), Glauber was recruited to assist in the creation of an atomic bomb in Los Alamos and help end World War II.

After arriving in New Mexico by train, Glauber shared a ride to the clandestine wartime lab with John von Neumann, a mathematician, physicist, and fellow child prodigy. The two would both make significant contributions to Fat Man — the plutonium, implosion-type weapon that would be released above Nagasaki, Japan on August 9, 1945.

For young Glauber, von Neumann was just one of many scientific greats at the Los Alamos laboratory — including Director J. Robert Oppenheimer and many future Nobel Prize winners — who would inspire him to pursue theoretical and optical physics after the war. Glauber went from working in the Theoretical Division at the laboratory to becoming a pioneer in the emerging field of quantum optics, ultimately earning the Nobel Prize in Physics in 2005.

Early years

Glauber was born in New York City in 1925 and spent his formative years traveling around the eastern United States with his salesman father and his mother. After moving to the Bronx, he became interested in astronomy and built telescopes and studied the night sky. Glauber joined the Junior Astronomy Club, which met at the Museum of Natural History in New York. He was accepted

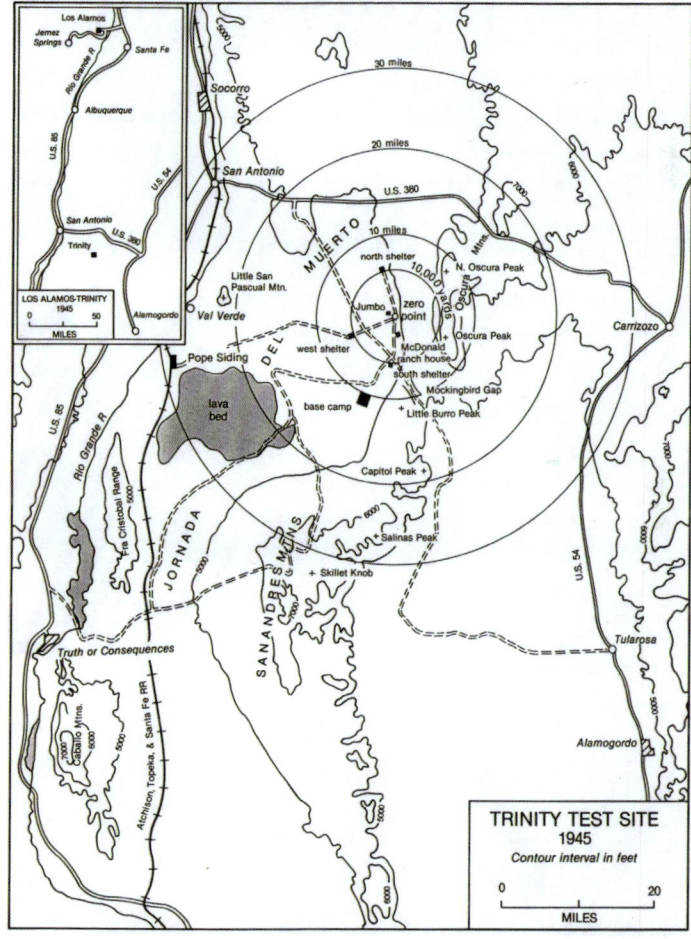

The first detonation of an atomic device was during the Trinity test. About 200-some miles south of the Los Alamos lab, Jornada del Muerto (often translated as Journey of the Dead Man) was selected as the Gadget's test site. (Public domain image.)

The Trinity test, which took place on July 16, 1945 in the New Mexico desert, was the world's first detonation of a nuclear weapon, marking the dawn of the Atomic Age.

A mostly uninhabited mesa in northern New Mexico became home to a top-secret lab and the community in which staff lived and their families lived. Los Alamos grew quickly and almost overnight.

to the prestigious Bronx High School of Science and spent his time learning calculus — an advanced area of study for the 1930s that wasn't usually taught in high schools — and focusing on experimentation for improving the spectroscope, used to measure the spectrum of light. He graduated from high school at the age of 16 and was accepted to Harvard University.

While Glauber was attending Harvard, the Japanese Navy Air Service forces attacked Pearl Harbor on December 7, 1941, thrusting the United States into World War II. When Glauber turned 18 in September 1943, he was required to register for the draft. He deferred because Harvard had hired him as a teacher, and he had signed up for the National Roster of Scientific and Specialized Personnel, which identified and recruited individuals for specific jobs in the war effort. Not long after registering, Glauber recalled years later in an oral history interview, that a man in a dark suit approached him about doing important work out West. Glauber also

recalled that until he had arrived in Los Alamos he did not realize that he would be contributing to the top-secret efforts to create nuclear weapons.

Los Alamos work

Glauber was first assigned to physicist Bruno Rossi's group, known as the RaLa Experiment Group, or Radioactive Lanthanum Experiment Group. In this group, Glauber studied the behavior of converging shock waves for the spherical implosion of plutonium. Glauber wanted to focus on theoretical physics, so he eventually made his way to the Theoretical Division, where he worked with Robert Serber. Serber was an engineer and physicist who was instrumental in explaining the basic principles and goals of the Manhattan Project, even printing lectures known as *The Los Alamos Primer*. Serber was a consultant in Project Alberta, which assisted in the delivery of the atomic bombs on Hiroshima and Nagasaki. Under Serber, Glauber helped calculate the bomb's critical mass, which is the amount of fissile material needed for a sustained nuclear reaction. This problem took almost two years to solve.

Los Alamos had an almost utopian feel for Glauber, and he relished in being among the nation's top scientific minds, including those who inspired Glauber's passion for theoretical and optical physics. Glauber felt the only real problem was the mud. With no paved sidewalks or roads, navigating the area was messy. Wooden walkways were eventually laid down as temporary sidewalks, but the hastily built town of Los Alamos had quirks that its residents had to cope with for the first few years.

As one of the younger members of the Manhattan Project and as a theorist, Glauber was not authorized to be at the Trinity test, which was in the New Mexico desert about 210 miles from Los Alamos.

There, on July 16, 1945, lab staff and members of the military would detonate The Gadget, which was an implosion plutonium-type weapon and precursor to the Fat Man released above Japan just weeks later. It was also the world's first detonation of a nuclear weapon, marking the beginning of the Atomic Age. Despite not being at the Trinity test site, Glauber and a few others watched from a distance at Sandia Peak in Albuquerque.

In later interviews, Glauber noted that he did not regret being a member of the Manhattan Project, however, the numerous delays during the 27 months it took to create the weapons and faster weapons development would have helped end the war sooner.

The hastily constructed town was often a muddy mess. Roy Glauber enjoyed living in Los Alamos, but the lack of paved sidewalks and roads was often problematic.

Post-World War II

Soon after World War II officially ended on September 2, 1945, Glauber returned to Harvard and later joined Oppenheimer at the Institute for Advanced Study in Princeton, New Jersey. Glauber taught for one year at the California Institute of Technology, replacing fellow Los Alamos scientist and future fellow Nobel laureate Richard Feynman (see Page 70), who was on sabbatical to Brazil. In 1952, not yet 30 years old, Glauber went back to Harvard, where he would spend the rest of his career as a faculty member in the physics department.

Glauber was intrigued by a particular experiment conducted in 1956: two British astronomers, Robert Hanbury Brown and Richard Twiss, attempted to measure the diameter of visible stars. They determined that light behaved as a continuous wave, and Glauber wanted to study the mathematics behind their discovery using quantum mechanics (the behavior of matter and light on the atomic and subatomic scale) applied to different forms of light. By developing a more sophisticated theory using quantum mechanics, Glauber was able to provide a better understanding of the behavior of light particles as both waves and a stream of particles, a conclusion that formed the basis for not just quantum optics, but also quantum computers, quantum cryptography, and even quantum mechanics.

Glauber's research continued with interactions of light with trapped ions and the fundamental nature of quantum jumps, which laid the groundwork for the fields of quantum cryptography and quantum computers. He received numerous awards for his research, and in 1997 he was elected as a foreign member of the Royal Society of London for Improving Natural Knowledge.

Roy Glauber arrived in Los Alamos to work at the clandestine lab when he was just 18 years old and a sophomore at Harvard University.

Glauber's research continued with interactions of light with trapped ions and the fundamental nature of quantum jumps, which laid the groundwork for the fields of quantum cryptography and quantum computers.

Glauber, Roy   J.   Harvard
Arrived: 1/27/44

with Bacher
with Serber

Last Day: Dec. 27, 1945

By 2005, he had earned the Nobel Prize in Physics for his findings on the quantum electrodynamic interactions of light and matter (Glauber was awarded half of the prize in 2005, sharing it with John L. Hall and Theodor W. Hansch who were each awarded one-fourth of the prize for their related work in laser-based precision spectroscopy and development of the optical frequency comb technique).

Glauber died in 2018 at the age of 93. He was survived by his children, Jeffrey and Valerie, and long-time companion.

Roy Glauber gave a director's colloquium at Los Alamos National Laboratory in 2015, nearly 70 years after his work as a staff member of the Manhattan Project.

ROY GLAUBER

1925–2018

LOS ALAMOS CONTRIBUTIONS

Theoretical Division, checking the calculations of the critical mass for the atomic bomb.

NOBEL PRIZE: PHYSICS, 2005

Studied the behavior of light particles that, like radio waves, can be considered both waves and a stream of particles. This concept explained the differences between diffuse light that scattered evenly, as is visible from a light bulb, and the intense beam of a laser or electromagnetic radiation. These differences illustrate a variety of frequencies versus a constant frequency and phase.

MARIA GOEPPERT MAYER

By Hadley Hershey

From an early age it was expected that Maria Goeppert Mayer

would continue the Goeppert family tradition of professorship.

Her father was a sixth-generation professor and she was to be the seventh.

However, universities' nepotism regulations, meant to discourage supervisors

from hiring relatives, denied employment to wives of professors and would

haunt Goeppert Mayer throughout her career.

For many years, each time her husband was offered a professorship, Goeppert Mayer was denied a similar position at the same university despite her formidable qualifications. Instead, she held volunteer positions, allowing the universities to benefit from her incredible scientific abilities, without having to pay her a salary.

Throughout her career, much of which was work in unpaid positions and that she said was "just for the fun of doing physics," Goeppert Mayer made contributions to the fields of nuclear physics, physical chemistry, and mathematics.

Some of her earliest work in her doctoral research presented the theory of two-photon absorption (2PA), which, after the invention of lasers in 1960, was experimentally confirmed. To honor Goeppert Mayer, the unit for 2PA cross-sections is called a Goeppert Mayer (GM) unit.

Near the end of her career, in 1963, she shared one-half of the Nobel Prize in Physics for her work on the nuclear shell model, which is the basis for understanding nuclear structure. Goeppert Mayer was just the second woman to be awarded the prize in physics (it would be another 55 years before the third). And in between these two feats, Goeppert Mayer contributed to the Los Alamos laboratory's efforts to create a nuclear weapon and help end World War II, ushering the world into the Atomic Age.

Early years

Goeppert Mayer was born on June 28, 1906 in Kattowitz, Germany (now Katowice, Poland), the only child of Friedrich Goeppert and Maria Wolff. In 1910, Friedrich became a professor of pediatrics at Göttingen University, and moved the family from Kattowitz to Göttingen. Goeppert Mayer attended public and private schools and entered Göttingen University in the fall of 1924. She began her studies in mathematics. At the request of Max Born, a family friend and professor of theoretical physics, she joined his quantum mechanics seminar. It was during this seminar that she discovered she preferred theoretical physics to mathematics, so she changed her course of study and became Born's doctoral student.

In 1928, Goeppert Mayer met Joseph Mayer, an American chemist on fellowship at Göttingen University. Joe arrived at the home Goeppert Mayer shared with her mother to inquire about renting a room. Because of the financial strain caused by Friedrich Goeppert's sudden death the year before, the family was renting rooms to students.

Joe and Maria were married January 18, 1930. In March, Goeppert Mayer received her Ph.D., and soon after Joe accepted a position in the chemistry department at Johns Hopkins University. The couple left Germany for Baltimore.

Because of the nepotism regulations, Goeppert Mayer was not offered a paid position at the university. Instead, she was a volunteer associate in the physics department until 1939, when the couple moved to Columbia University in New York. Nepotism rules would again keep Goeppert Mayer from a paid position there. At the request of Harold Urey, professor of chemistry, she was a volunteer lecturer in the chemistry department.

In 1941, Goeppert Mayer was offered her first paid job: a substitute position at Sarah Lawrence College. She taught mathematics that first year, and in June 1942 she was reappointed to the science faculty. During the 1942-1943 academic year, Goeppert Mayer taught four courses: two in mathematics, one in physics, and one in physical chemistry. She was also still a volunteer chemistry lecturer at Columbia University.

Los Alamos work

In September 1943, Urey, now the director of the Substitute Alloy Materials (SAM) Laboratories at Columbia, needed an expert in spectral analysis to lead a team researching the

photochemical method for uranium isotope separation. Urey was familiar with Goeppert Mayer's earlier work and knew she was the right person for the job. He wrote to the president of Sarah Lawrence College, requesting a one-year leave of absence for Goeppert Mayer to return to Columbia full time for "an important war project."

When it came to deciding who should lead the Opacity Project, Goeppert Mayer was an obvious choice.

The Otowi Bridge was built above the Rio Grande in 1924 and was important to the lab's early transportation needs.

The SAM Laboratories were conducting research for the Manhattan Project, which was the U.S.-led effort to create the world's first atomic weapons. The focus of the wartime research at Columbia was looking into the uranium isotope separation processes required to produce the weapons-grade uranium desperately needed by the scientists at the clandestine laboratory in Los Alamos.

Urey's request was granted, and Goeppert Mayer returned to Columbia full time. Goeppert Mayer supervised a team at the SAM Laboratories that was researching the photochemical method for uranium isotope separation. This work involved analyzing the spectra of uranium-235 and uranium-238. Within a few months, the photochemical method was deemed impractical and was abandoned. Goeppert Mayer was moved to the team researching gaseous diffusion. Gaseous diffusion and electromagnetic separation were deemed more practical, given the time constraints of producing a weapon before the Nazis had their own.

Meanwhile, in late May of 1944, Los Alamos laboratory Director J. Robert Oppenheimer and physicist Edward Teller needed a group that could undertake opacity calculations, which dealt with the properties of radiation at high temperatures. Oppenheimer asked General Leslie Groves, who was leading the Manhattan Project and overseeing the secret work in Los Alamos, to allow the work to take place at the SAM Laboratories at Columbia. When it came to deciding who should lead the Opacity Project, Goeppert Mayer was an obvious choice. She was already involved with the Manhattan Project efforts and was a skilled mathematician and theorist, head of a team of researchers, and an expert in spectral analysis.

With Groves' approval, Goeppert Mayer was hired as a consultant for Los Alamos. Teller would oversee the project from Los Alamos,

as his frequent travels to New York for other Manhattan Project business would allow him to meet with Goeppert Mayer at Columbia without raising suspicions.

At the SAM Laboratories, Goeppert Mayer enlisted the help of two of her graduate students. The three began work on the opacity

Little Boy was a uranium, gun-type weapon and the first atomic bomb to be used in combat. Maria Goeppert Mayer's wartime research focused on aspects of uranium that would affect weapons development. (Public domain image.)

Maria Goeppert Mayer's inprocessing personnel card noted her arrival in Los Alamos on May 21, 1945 as a visitor of physicist Edward Teller. She was hired as a consultant to the Los Alamos lab, though most of her war-related work was conducted at Columbia University.

Mayer, Maria
Arrival: 5/2Ø/45

with Teller, visitor

calculations and were in regular communication with Teller. On May 21, 1945, Goeppert Mayer arrived in Los Alamos to continue the opacity calculations with Teller. While not directly beneficial to the development of fission weapons at Los Alamos during the war, the calculations were later applied to the development of the hydrogen bomb.

Post-World War II

In 1946, Goeppert Mayer and her husband moved to Chicago. Joe was appointed professor in the chemistry department at the Institute for Nuclear Studies at the University of Chicago. Despite her work for the Manhattan Project, Goeppert Mayer was once again denied a paid position and worked for free as a professor of physics at the institute.

In July, Goeppert Mayer was offered the position of senior physicist at the newly formed Argonne National Laboratory in Chicago. She continued her unpaid professorship at the University of Chicago, while working part time at Argonne. It was during her time at Argonne that Goeppert Mayer would develop the nuclear shell model that would later earn her the Nobel Prize in Physics.

Goeppert Mayer was working on a project with Teller to determine the origin of elements. They were looking at isotope abundances when Goeppert Mayer observed patterns. She noticed that nuclei with 2, 8, 20, 28, 50, 82, or 126 protons or neutrons were stable (known as the magic numbers). Goeppert Mayer recognized the importance of these numbers and speculated they could help describe the structure of the nucleus, and perhaps nuclei had a shell structure similar to the electron shell structure in atoms. Goeppert Mayer was not the first to notice this phenomenon, which was studied in the 1930s, so she reviewed the earlier research and continued to study the data.

By 1948, she had the data but not the theoretical explanation for the shell structure of nuclei. She presented her data in a paper "On Closed Shells in Nuclei," published in the *Physical Review* journal that same year. Early the next year, during one of many discussions of shell structure with physicist Enrico Fermi (see Page 64), he remarked, "Is there any indication of spin-orbit coupling?" This was the breakthrough Goeppert Mayer needed. She now had her theoretical explanation for shell structure, and the spin-orbit-coupling shell model of nuclei was presented in the February 4, 1949 issue of the *Physical Review*. Around this same time J. Hans D. Jensen, a German physicist, came to the same conclusions as Goeppert Mayer. The two met in 1951 and started a collaboration that culminated in 1955 with the publication of their book, *Elementary Theory of Nuclear Shell Structure*.

In 1960, Goeppert Mayer accepted a full professorship in physics at the University of California at San Diego. It was a paid position. Soon after arriving in San Diego, Maria had a stroke at the age of 53, which would affect her health for the remainder of her life. However, she continued to teach and contribute to the field of physics.

In 1963, Goeppert Mayer was awarded half of the Nobel Prize in Physics, sharing it with Jensen for their discoveries concerning nuclear shell structure. The other half of the prize that year was awarded to Eugene Paul Wigner for unrelated work.

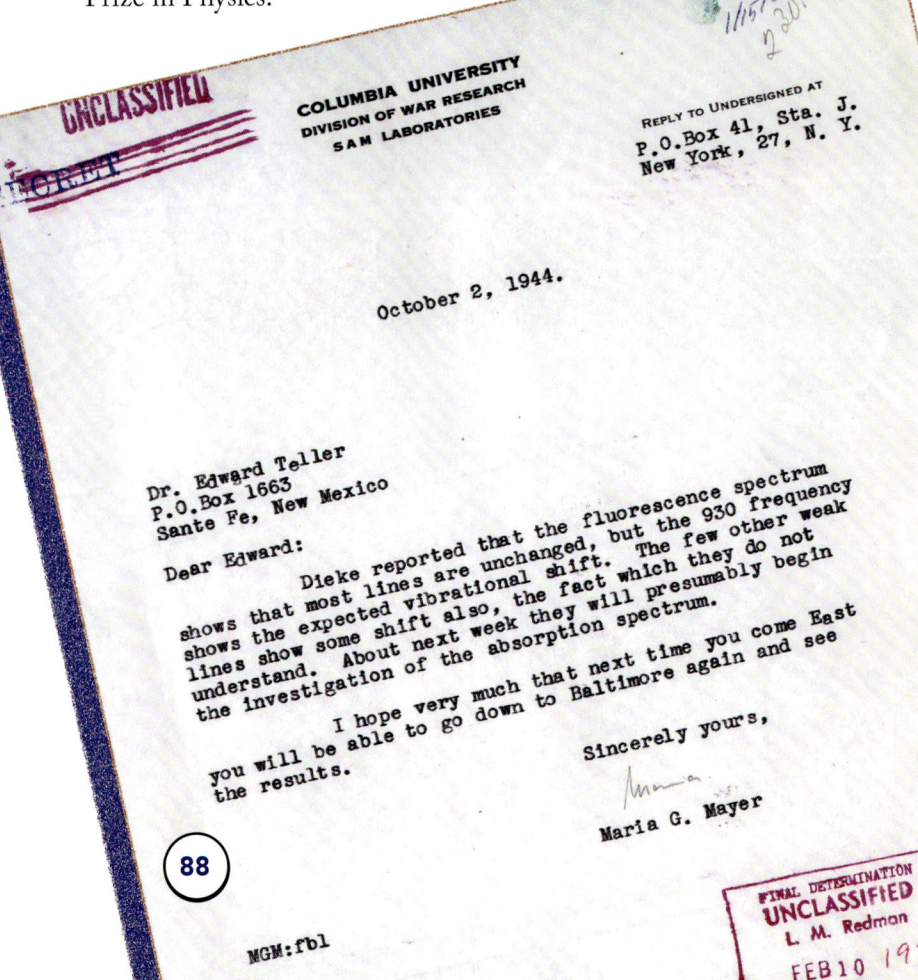

This 1944 letter from Maria Goeppert Mayer to physicist Edward Teller, along with photos and other historic materials from the Manhattan Project era, are part of the National Security Research Center's unclassified legacy collections.

Goeppert Mayer died on February 20, 1972, after suffering from a massive heart attack and then spending weeks in a coma. She was survived by her husband, daughter Maria Ann, and son Peter Conrad. According to the *Scientific American*, before her death Goeppert Mayer told a group of high school girls, "Become fully educated women and promote the understanding of science in any way you can. Our country needs your help. My generation has played its part. It is up to you to carry on."

Maria Goeppert Mayer was a recipient of the 1963 Nobel Prize in Physics. She contributed to work at the wartime lab in Los Alamos, as staff raced to develop the atomic bomb and help end World War II. (Photo courtesy of the U.S. Department of Energy.)

MARIA GOEPPERT MAYER
1906–1972

LOS ALAMOS CONTRIBUTIONS

Supervised project on photochemical method for uranium isotope separation at SAM Laboratories, 1943. Supervised Opacity Project at SAM Laboratories, 1944. Worked with physicist Edward Teller on the Opacity Project at Los Alamos, 1945.

NOBEL PRIZE: PHYSICS, 1963

Developed the nuclear shell model, which describes the structure of the nucleus.

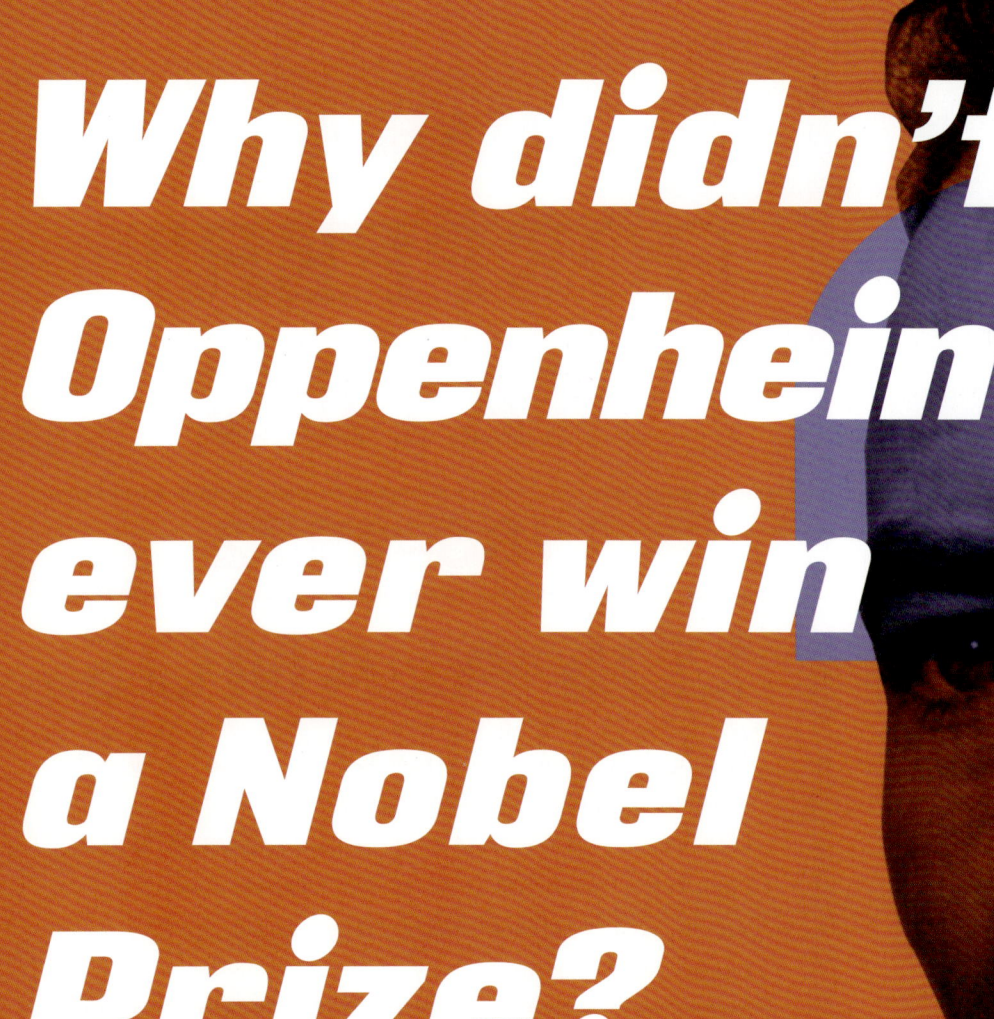

Why didn't Oppenheimer ever win a Nobel Prize?

BY BRYE STEEVES

For his scientific achievement, he would be forever known as the father of the atomic bomb — but never as a Nobel laureate.

The pinnacle of global recognition, the Nobel Prize was awarded to 18 of J. Robert Oppenheimer's colleagues with whom he worked at the Manhattan Project site in Los Alamos. There, in just 27 months and in a perceived race with Nazi Germany, the scientists created the first nuclear weapons. Their efforts brought the world into the Atomic Age and helped end World War II. Several earned the prize before coming to work at the wartime lab, while most would go on to win later in life.

Oppenheimer was nominated for the Nobel Prize for Physics three times: in 1946, in 1951, and in 1967. Colleagues, scholars, and surely Oppenheimer himself pondered why he was never bestowed the honor.

"To understand this," said James Kunetka, historian and author of *The General and the Genius*, "you have to first examine the man's academic life before and after the war."

Undisputed genius

Born in 1904 into a wealthy Jewish family and raised in New York, Oppenheimer was obviously gifted. He completed both third and fourth grades in just one year and later skipped a portion of his eighth-grade year. Remarkable anecdotes of brilliance illustrate his life through early adulthood. As a boy, he was interested in mineralogy and, at age 12, presented his research paper to the New York Mineralogical Club and became an honorary member. As a young academic, he learned Dutch in six weeks to successfully deliver a technical lecture on a trip to the Netherlands. It was there he was first dubbed "Oppie" ("Opje" in Dutch).

"[He was] one of the sharpest people I have ever seen or heard of, intellectually," said longtime friend Harold Cherniss in a 1979 interview. "When he became interested in anything, he very quickly picked up an enormous amount of knowledge about it."

After graduating at the top of his high school class, Oppenheimer studied science at Harvard University, where he was admitted to graduate-level physics classes during his first year. He also took courses in literature, languages, religion, and philosophy, earning his degree in just three years, but with no social clubs or athletics listed under his name in the 1925 yearbook. Certainly introverted, but also perhaps lonely, Oppenheimer once told a friend, "It's no fun to turn the pages of a book and say, 'Yes, yes, of course, I know that,'" according to an October 1949 article in *Life* magazine.

After a stint at the University of Cambridge in the United Kingdom, Oppenheimer went to the University of Göttingen in Germany, where he studied quantum physics and earned his doctorate in 1927. By 1929, had accepted offers to teach at both the California Institute of Technology (Caltech) and the University of California, Berkeley.

Oppenheimer's early research focused on energy processes of subatomic particles, including electrons, positrons, and cosmic rays, as well as neutron stars (collapsed cores of massive stars) and black holes. He was soon recognized as a leader in theoretical physics and had earned the respect of scientific greats, such as Albert Einstein and Niels Bohr (see Page 42).

"However, many of his colleagues and critics point out that his production of significant papers was surprisingly thin," Kunetka said. "It was said by some that he far too often co-authored papers with his students rather than initiated them. Hans Bethe (see Page 22) noted that while Oppenheimer and others were perhaps more brilliant, he (Bethe) was more productive."

Oppenheimer proved to be an outstanding teacher, inspiring and influencing students. He earned a loyal following, if not outright adoration. "Like most of his students, I would more or less follow him to the ends of the earth," recalled Manhattan Project scientist Robert Christy in a 1983 interview.

Leadership

Oppenheimer lacked large-scale managerial experience and his associations with Communist party members were problematic. And without a Nobel Prize, it wasn't certain whether Oppenheimer would have the prestige to direct the Los Alamos scientists.

But as soon as General Leslie Groves met Oppenheimer, none of that mattered, according to an Oppenheimer biographer Ray Monk. The young professor impressed the Manhattan Project leader with both his intelligence and practicality. Oppenheimer would seemingly be able to turn blackboard theories and lab experiments into atomic weapons. Groves also may have seen a drive-based ambition, according to Monk, assuring him that Oppenheimer would, indeed, succeed.

In the fall of 1942, Groves hired the 38-year-old Oppenheimer, who recommended Los Alamos as the site for the clandestine lab and recruited science's greatest minds to join him there. By then, Oppenheimer was described as charismatic and charming. He was the center of attention at parties, drinking his signature martinis and gesticulating with cigarettes through story after story, his bright blue eyes sparkling.

"Oppenheimer commanded not just the loyalty but the deep respect of everybody who was at Los Alamos, and I cannot think of anyone else who would have succeeded as he did in that sense," said Manhattan Project physicist and Nobel laureate Roy Glauber (see Page 84).

But Oppenheimer was also known as cruel and intolerant, and was called a showman and a power seeker. Even still,

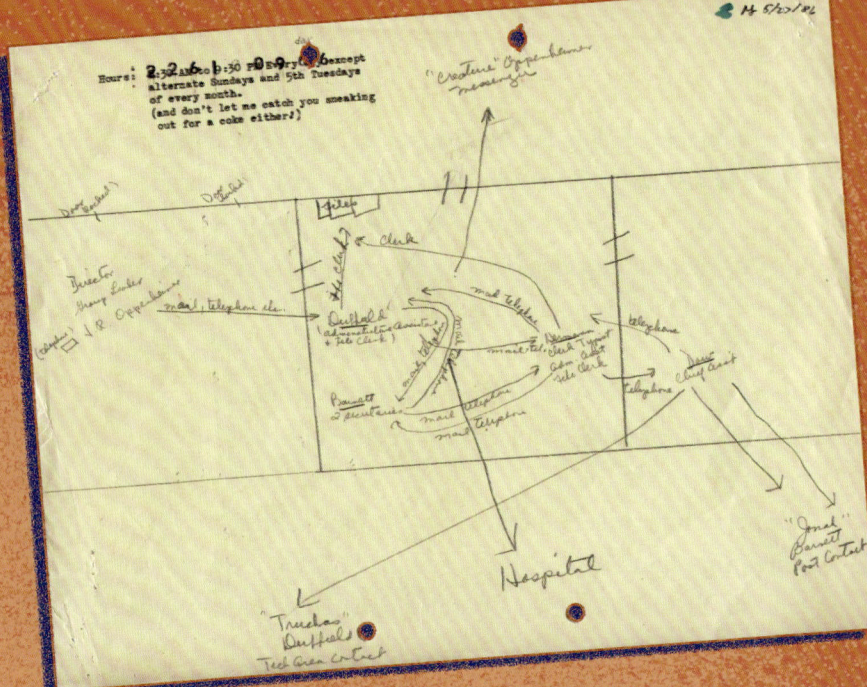

This sketch is the only known imagery of Lab Director J. Robert Oppenheimer's wartime office at the Los Alamos laboratory. The lab also has his office chair, which is often on display in its public museum.

said Alan Carr, senior historian at Los Alamos National Laboratory's National Security Research Center, Oppenheimer's worst enemies would concede that he achieved greatness during the war.

"He was very close to being indispensable," an unnamed Los Alamos scientist said, according to the 1949 *Life* article. Another said, "[t]he main decisions were made by Oppenheimer, and all proved to be correct."

Atomic success

Oppenheimer's directorship, perhaps along with his genius, culminated on July 16, 1945 when the world's first-ever atomic device was successfully detonated in the New Mexico desert. Oppenheimer, who read and wrote poetry, named the test "Trinity." He said years afterward that he may have been inspired by a John Donne poem that includes the line: "Batter my heart, three-person'd God."

Weeks later and just days apart, the United States released the gun-type uranium bomb, Little Boy, and the implosion-style plutonium bomb, Fat Man, above Japan. Groves phoned Oppenheimer after the first detonation.

According to a transcript of the recorded call, Groves said, "I think one of the wisest things I ever did was when I selected [you] the director of Los Alamos."

To which Oppenheimer responded: "Well, I have my doubts, General Groves."

And Groves replied: "Well, you know I've never concurred with those doubts at any time."

After World War II

Oppenheimer once said "physics and desert country" were his "two great loves." It was in Los Alamos that these came together, and where his work as a physicist changed the world.

He left Los Alamos a few weeks after the official end of World War II on September 2, 1945. "Oppenheimer" was by then a household name. With his face on magazine covers, star-treatment followed. His celebrity, though, did not translate into a Nobel Prize.

He first returned to Caltech, but soon left to lead the Institute for Advanced Study in Princeton, New Jersey, as well as serve as the chairman of the General Advisory Committee,

J. Robert Oppenheimer's badge photo as the lab's first director is part of the collections of the National Security Research Center.

which was a scientific panel that advised the newly formed Atomic Energy Commission. Much of his focus shifted from his prewar physics to policy work. Oppenheimer spoke out in opposition to the development of the even more powerful hydrogen bomb, questioning its feasibility early on, but also deeming it an unnecessary weapon. Meanwhile, he wrote and lectured but did not, however, resume much research.

In 1954, he lost his security clearance following unsubstantiated accusations against his loyalty. Though his supporters remained steadfast and numerous, Oppenheimer eventually retreated from his public life and work, pushing him further than ever from a Nobel Prize.

Why no prize?

Kunetka says that the simplest explanation is that before World War II, Oppenheimer's published work was not considered significant enough. Carr agrees, adding that Oppenheimer never made a major discovery, nor did he ever prove a significant theory.

"The Nobel Prize requires more than just a remarkable idea," Carr said, "it requires evidence."

For his Manhattan Project work, Oppenheimer himself said that creating the atomic bomb was inventive rather than scientific, according to the *Life* 1949 article.

When he was first nominated in 1946 for the Nobel Prize, the Nobel committee was hesitant to award it to someone so closely tied to the atomic bomb, according to *American Prometheus*. Many scholars and scientists through the years have concurred.

Others, though, have said Oppenheimer's scientific focus would change frequently and he didn't work sufficiently in any one area to warrant the Nobel Prize. Meanwhile, Monk thought Oppenheimer's work was more significant than credited and some scientists, including Nobel laureate Luis Alvarez (see Page 14), speculated that Oppenheimer's work on black holes may have warranted the Nobel had he lived long enough to see them brought to fruition (the prizes are not awarded posthumously).

"In the end," Kunetka said, "we don't know." Carr added, "Did he achieve greatness? Yes, of course. What Oppie led his wartime team of scientists to achieve was nothing short of remarkable. He will always have that incredible scientific legacy."

Loyal following until the end

Oppenheimer died at his New Jersey home in 1967 after unsuccessful treatments for throat cancer. He was 62 and was survived by his wife, Kitty, and their two children, Peter and Katherine. Kitty spread his ashes near their simple beach home in the U.S. Virgin Islands, following a memorial service at Princeton University. An estimated 600 people attended.

"Science is not everything," Oppenheimer once said, "but science is very beautiful."

EDWIN MCMILLAN

By Renae Mitchell

From a young age, Edwin McMillan was obsessed with unraveling nature's secrets while concurrently enjoying its complex beauty. It was this fixation that enabled this chemist and physicist to make various scientific contributions during his lifetime. These achievements include discovering the chemical element neptunium and the isotopes plutonium-239 and oxygen-15; developing the gun-type design used for the uranium bomb; contributing to the implosion-assembly concept used for the plutonium bomb; and determining the principle of phase stability used for nuclear accelerators to produce beams of charged particles for research purposes.

When not making contributions to science, McMillan could be found exploring the Anza-Borrego desert in Southern California, where he enjoyed collecting specimens for his rock garden or swimming in the Pacific Ocean, where he collected shells. He also enjoyed gardening and, in his earlier years, climbed mountains, including California's Mount Whitney and the Matterhorn in the Alps.

Colleagues described McMillan as modest and a man of few words. And they also noted his wit, which was inspired by the stylized humor of cartoonist James Thurber, whose characters escape into fantasy because they are befuddled by a world they neither created nor understood, and the sharp yet bizarre comedy of the Marx brothers.

"He was incurably curious," one of his grown sons, Stephen McMillan, noted. "He was an explorer. He wasn't looking at what was known but what wasn't."

Edwin McMillan helped J. Robert Oppenheimer, director of the Los Alamos lab, select a suitable site to develop what would be the world's first atomic device. McMillan and four others chose an isolated mesa in northern New Mexico.

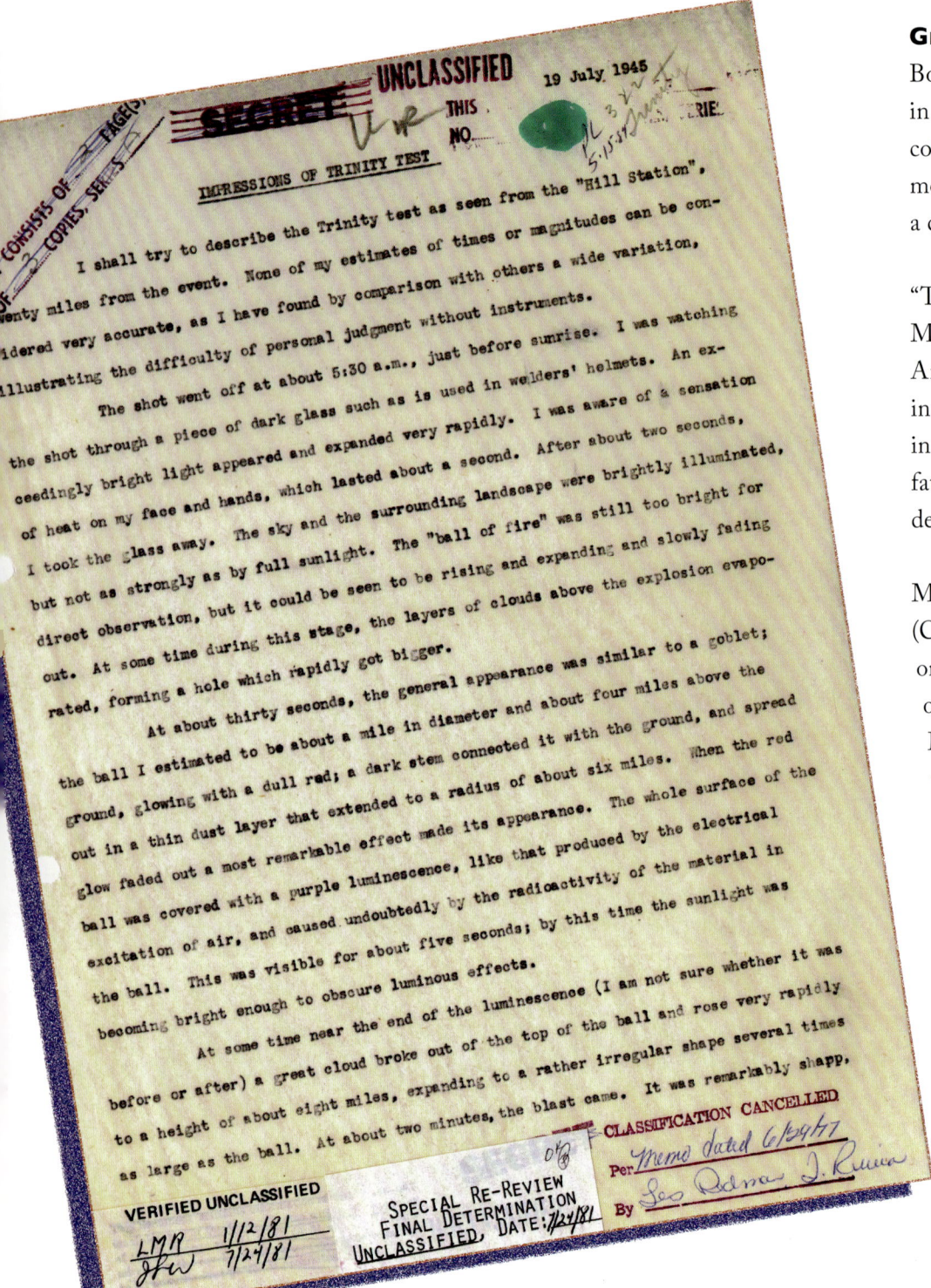

UNCLASSIFIED
SECRET
19 July 1945

THIS NO.

CONSISTS OF PAGE(S)
COPIES, SERIES

IMPRESSIONS OF TRINITY TEST

I shall try to describe the Trinity test as seen from the "Hill Station", twenty miles from the event. None of my estimates of times or magnitudes can be considered very accurate, as I have found by comparison with others a wide variation, illustrating the difficulty of personal judgment without instruments.

The shot went off at about 5:30 a.m., just before sunrise. I was watching the shot through a piece of dark glass such as is used in welders' helmets. An exceedingly bright light appeared and expanded very rapidly. I was aware of a sensation of heat on my face and hands, which lasted about a second. After about two seconds, I took the glass away. The sky and the surrounding landscape were brightly illuminated, but not as strongly as by full sunlight. The "ball of fire" was still too bright for direct observation, but it could be seen to be rising and expanding and slowly fading out. At some time during this stage, the layers of clouds above the explosion evaporated, forming a hole which rapidly got bigger.

At about thirty seconds, the general appearance was similar to a goblet; the ball I estimated to be about a mile in diameter and about four miles above the ground, glowing with a dull red; a dark stem connected it with the ground, and spread out in a thin dust layer that extended to a radius of about six miles. When the red glow faded out a most remarkable effect made its appearance. The whole surface of the ball was covered with a purple luminescence, like that produced by the electrical excitation of air, and caused undoubtedly by the radioactivity of the material in the ball. This was visible for about five seconds; by this time the sunlight was becoming bright enough to obscure luminous effects.

At some time near the end of the luminescence (I am not sure whether it was before or after) a great cloud broke out of the top of the ball and rose very rapidly to a height of about eight miles, expanding to a rather irregular shape several times as large as the ball. At about two minutes, the blast came. It was remarkably sharp,

CLASSIFICATION CANCELLED
Per *memo dated 6/29/77*
By

VERIFIED UNCLASSIFIED
LMR 1/12/81
JFW 7/24/81

SPECIAL RE-REVIEW
FINAL DETERMINATION
UNCLASSIFIED, DATE: 7/24/81

Edwin McMillan documented his impressions of the Trinity test, including its brilliant colors and illumination of the surrounding desert landscape.

Growing up on the West Coast

Born in 1907 in Redondo Beach, California, McMillan grew up in nearby Pasadena, where he enjoyed tinkering with gadgets and collecting rocks and minerals. McMillan was surrounded by family members who worked in medicine, including his father, who was a doctor.

"Three of [my] mother's brothers were physicians in Pasadena," McMillan said in a 1972 oral history interview recorded by the American Institute of Physics. "My father's twin brother practiced in Los Angeles. Also, two of my father's many brothers-in-law, one in Los Angeles and one in Accident, Maryland. Both mother and father came from large families, where there were also a couple of dentists. It was a very medically oriented family."

McMillan attended the California Institute of Technology (Caltech), earning a bachelor's degree in 1928 and a master's degree one year later. He then attended Princeton University, where he obtained a Ph.D. in 1932. Although his degrees were in physics, McMillan delved heavily into chemistry. Indeed, his first paper was a chemical study, titled "An x-ray study of the alloys of lead and thallium," which was published in 1927 in the *Journal of the American Chemical Society*.

After completing his doctorate, McMillan joined the University of California, Berkeley, as a National Research Fellow and then joined Berkeley's Radiation Laboratory. There, he worked under Nobel laureate and professor Ernest O. Lawrence to study nuclear reactions and their products. McMillan was appointed as an instructor at the university in 1935. A year later, he became an assistant professor; in 1941, he was appointed as a full professor.

Oppenheimer comes calling

During the late 1930s, McMillan was busy building on the work he had begun with Emilio Segrè (see Page 136). Along with a former student turned collaborator, Philip Abelson, McMillan suspected the existence of a new element but could not yet definitively prove its existence. Their experiments centered on a radioactive source that emerged from the decay of uranium, one that was distinct from all known chemical elements. They named this decay product (the radioactive atom resulting from the atomic fission of uranium) "neptunium" after the planet Neptune.

Through a paper published in the journal *Physical Review*, McMillan and Abelson introduced the newly discovered neptunium. The radioactive metal was the first transuranic element discovered. Transuranic elements have atomic numbers greater than 92, meaning that they lie beyond uranium (named after the planet Uranus) on the periodic table.

McMillan had just co-discovered neptunium in 1940 and then plutonium-239 in 1941, when none other than the soon-to-be director of a secret wartime lab, J. Robert Oppenheimer, came calling.

Oppenheimer asked McMillan to help him select a site that would be used as a base of operations to develop what would be the world's first atomic weapon. McMillan, along with four others, helped select an isolated mesa in Los Alamos, New Mexico, as the most suitable location for the lab of the Manhattan Project (the U.S. government's top-secret effort to build the first atomic bombs to help end World War II). McMillan, Oppenheimer, Lawrence, General Leslie Groves, and Colonel John H. Dudley first looked at Jemez Springs, but Groves disapproved. Oppenheimer then suggested that they look at the Los Alamos Ranch School on a nearby mesa. With Groves's approval, McMillan, Oppenheimer, and Lawrence then returned to Los Alamos on a separate trip to verify that it was the best location available.

Edwin McMillan helped show the plutonium gun-type weapon called Thin Man was not technically feasible.

From 1942 to 1945, McMillan was critical to the formation of what became the Los Alamos wartime lab. Playing a crucial role in the development of the uranium-gun assembly weapon called Little Boy, McMillan also helped to demonstrate that the plutonium gun-type weapon (Thin Man) was not technically feasible. He advocated that resources were better spent developing a plutonium implosion-style weapon. By working in the Gadget Division, McMillan contributed to developments in implosion, which in part led to the development of the Fat Man bomb.

The Little Boy and Fat Man atomic weapons were released above Japan on August 6 and 9, 1945, respectively, helping to end World War II on September 2, 1945.

After the war, McMillan went on to co-discover the principle of phase stability, which helped lead to the development of the synchrotron and synchro-cyclotron, which are machines used to examine the molecular and atomic details of a variety of materials. He was then co-awarded the Nobel Prize in Chemistry in 1951, sharing it with Glenn T. Seaborg, who discovered the second known transuranic element — plutonium — for "discoveries in the chemistry of the transuranium elements."

However, the Nobel Prize was not the only significant honor McMillan received. He was elected to the National Academy of Sciences in 1947, received the Research Corporation Scientific Award in 1951, and won the Atoms for Peace Award in 1963. In 1990, McMillan received the science community's highest civilian honor in the United States: the National Medal of Science, presented to him by President George Bush.

Final decades at the Berkeley Radiation Laboratory

McMillan became deputy director of the Berkeley Radiation Laboratory in 1954. In August 1958, he became director of the laboratory, a position he held for 15 years until his retirement in 1973. He continued to participate in research at the laboratory until his health severely declined in 1984 after suffering a number of strokes.

"His important and versatile scientific contributions, spanning physics, chemistry, and engineering, and his great human qualities, form an important chapter in the history of science."

Edwin McMillan left his teaching position in California to join the Manhattan Project. He was critical to the wartime lab's development.

He died in 1991 and was survived by his wife Elsie Blumer and their three children, Ann, David, and Stephen. Elsie reportedly saw from the window of her cottage the flash of the Trinity test, detonated in the early morning hours to help keep it secret. Four years after her husband's death, she published a book about her life in Los Alamos titled *The Atom and Eve*.

McMillan's co-recipient for the Nobel Prize, Glenn T. Seaborg, said in McMillan's obituary in *The New York Times*, "His important and versatile scientific contributions, spanning physics, chemistry and engineering, and his great human qualities, form an important chapter in the history of science."

Others also fondly remembered McMillan, pipe in mouth, cruising along the West Coast in his top-down 1957 Ford Thunderbird.

Edwin McMillan, shown here in his Los Alamos badge photo, left the lab after World War II ended. He was awarded the Nobel Prize in Chemistry in 1951.

EDWIN MCMILLAN

1907–1991

LOS ALAMOS CONTRIBUTIONS

Assisted J. Robert Oppenheimer in constructing the Little Boy atomic bomb that was dropped on Hiroshima, Japan, on August 6, 1945.

NOBEL PRIZE: CHEMISTRY, 1951

Awarded with Glenn T. Seaborg for the discovery of plutonium and neptunium.

ISIDOR RABI

By Octavio Ramos, Jr.

Laymen likely consider him "a creature scattering antibiotics with one hand and atomic bombs with the other," Isidor Rabi once said, according to _The New York Times_.

Described as a "humanistic scientist" in a 1964 story in _The Times_, Rabi earned a reputation for what fellow scientists called his "street smarts" when it came to atomic physics — not too surprising, since Rabi spent some time growing up on the streets of New York's Lower East Side, where his father worked as a tailor. Nobel laureate Hans Bethe (see Page 22) further defined Rabi's street smarts by noting that "Rabi always found a simpler way to do any given experiment, and this made him a great physicist."

Although scientists admired Rabi's tenacity, they also regarded him as the "conscience of their community," in particular for his moral influence later in the debate over harnessing the power of the atom exclusively for peaceful purposes. This is not to say that Rabi was not averse to using technology for the war effort. In a 1988 story in _The New York Times_, Rabi explained that he had stood ready to support any idea that could help defeat Adolf Hitler during World War II. Rabi remembered, "Some person would come along with a bright idea and I'd say, 'How many Germans will it kill?'"

In a June 1970 letter written to his friend and longtime associate Jerrold Zacharias, Rabi articulated his definition of science: "To me, science is an expression of the human spirit, which reaches every sphere of human culture. It gives an aim and meaning to existence as well as knowledge, understanding, love, and admiration for the world. It gives a deeper meaning to morality and another dimension to esthetics."

Early years
Born in Raymanov, then Austria-Hungary and what is now Poland, on July 29, 1898, Rabi came to the United States as a baby and was raised in New York. Rabi acquired his love of science when he was a

Isidor Rabi (right) converses with nuclear scientist Ernest O. Lawrence (left) and physicist Enrico Fermi (center). The three men contributed to various aspects of the Manhattan Project, which was the U.S. government effort to secretly create the atomic bomb.

boy, learning from books borrowed from the public library to build, among other things, his own radio set. Astonishingly, he published his first scientific paper on electronics in *Modern Electrics* while still in elementary school.

Rabi received a bachelor's degree in chemistry from Cornell University in June 1919, but because at the time Jews were excluded from employment in industry and academia, he spent three long years achieving little, if anything, professionally. He decided to return to Cornell, where he soon switched his graduate studies from chemistry to physics because the former no longer captivated him. He then transferred to Columbia University, where in 1927 he received a Ph.D. in physics for work on the magnetic properties of crystals.

In early 1929, Rabi started lecturing at Columbia, but he quickly earned a poor reputation. Physical chemist Harold Urey described Rabi as one of "the worst teachers I ever had." Despite this, Rabi became a full professor in 1937.

Part of what enabled Rabi's professorship was a shift in 1931 from teaching to conducting particle beam experiments. After developing the Breit-Rabi equation with collaborator Gregory Breit, Rabi went on to establish a molecular beam laboratory at Columbia, which soon attracted various scientists interested in using the instrument for various experiments. Such work culminated in the discovery of the magnetic resonance method in 1938.

Months after the discovery of nuclear fission in Germany in December 1938, fellow physicist and Nobel laureate Enrico Fermi (see Page 64) and Rabi were in Washington, D.C., to participate in a conference on theoretical physics. While at the conference, the two men met with Niels Bohr, a physicist who won the Nobel Prize in 1922 (see Page 42), to discuss the topic of nuclear fission.

It is estimated that more than 600,000 people worked on the Manhattan Project. There were three main sites: Los Alamos, New Mexico; Hanford, Washingtron; and Oak Ridge, Tennessee.

This meeting was Rabi's first taste of nuclear fission and its potential ability to produce incredible energies. Six years later, Rabi would experience these energies firsthand at the Trinity test held at the Jornada del Muerto desert in New Mexico the summer of 1945.

Helping sculpt Los Alamos

Rabi met J. Robert Oppenheimer, the future "father of the atomic bomb," in 1929. It seemed that Oppenheimer had kept an eye on the physicist, especially Rabi's successful oversight of the Radiation Lab at the Massachusetts Institute of Technology. In 1942, Oppenheimer reached out to Rabi, offering him the position of deputy director in charge of experimental work at a new top-secret project. As Rabi put it, he refused to ever "go on the payroll" for what would become the Manhattan Project, which included the U.S. government's top-secret lab in Los Alamos to create the first atomic bombs. Instead, he agreed to serve as Oppenheimer's personal advisor.

Along with Manhattan Project physicist Robert Bacher, Rabi advised Oppenheimer to change his plan to establish a military-led laboratory in New Mexico and instead form a team led by civilian scientists. Oppenheimer took the advice, and as a result this new laboratory would be run by the University of California under contract with the U.S. War Department.

According to Bethe, Rabi was responsible for giving the laboratory its initial organizational structure, one that exists to the present day. "Without Rabi, it would have been a mess because Oppie did not want to have an organization," Bethe said. "Rabi came to Oppie and said, 'You have to have an organization. The laboratory has to be organized into divisions and the divisions into groups. Otherwise, nothing will ever come of it.' And, Oppie, well, that was all new to him. Rabi made Oppie more practical."

Throughout Project Y (the codename for the clandestine lab) of the Manhattan Project, Rabi served as Oppenheimer's principal troubleshooter, finding ways to recruit prominent scientists like Bethe and Bacher, and advising Oppenheimer on political pitfalls

By 1942, J. Robert Oppenheimer, the first director of the Los Alamos lab, asked Isidor Rabi to join the effort to create the first atomic bombs. Rabi agreed to work as Oppenheimer's advisor.

that ranged from squabbles among the scientists at Los Alamos to difficulties with the central Europeans. Rabi also served on Project Y's technical board, where he addressed issues associated with physics and technique.

On July 16, 1945, Rabi attended Trinity, the codename given for the world's first detonation of a nuclear device. Wearing welding goggles, Rabi stood next to Fermi during what Rabi called the "longest 10 seconds of his life."

Rabi recounted: "Suddenly, there was an enormous flash of light, the brightest light I have ever seen or that I think anyone else has ever seen. It blasted; it pounced; it bored its way into you. It was a vision which was seen with more than the eye."

Rabi's loyalty to Oppenheimer never wavered. He testified at Oppenheimer's hearing before the Personnel Security Board on May 5,

1954. The hearing was prompted by questions of Oppenheimer's loyalty, and although the allegations were never substantiated, Oppenheimer's security clearance was revoked, prompting outrage. At the hearing, Rabi quipped, "We have an A-bomb ... What more do you want, mermaids?"

Always a New Yorker at heart

Rabi received the Nobel Prize in Physics in 1944 for his work in developing a magnetic resonance method capable of observing atomic spectra. This method enables scientists to measure the magnetic properties of atoms, atomic nuclei, and molecules. In an ironic twist of fate, physicians used magnetic resonance imaging (MRI) — developed from his work on magnetic resonance — to examine him during the last days of his life many decades later.

Following the end of World War II in 1945, Rabi spent much of his remaining life as an advisor. For example, he played a role in working with European scientists in establishing CERN, the largest particle physics laboratory in the world. He also served as President Dwight Eisenhower's science advisor and acted as the U.S. representative to the NATO Science Committee.

Although notably stubborn, Rabi earned a reputation as a wise and diplomatic leader, even from an early age. This was reinforced by both Edward Purcell at the Radiation Laboratory and Richard Feynman (see Page 70) at Los Alamos, both of whom called him an elder statesman, even though he was still a young man at the time. His influence in the world of physics led many to call him "the dean of world physics."

In addition to winning the Nobel Prize, Rabi received countless awards, among them the Elliott Cresson Medal from the Franklin Institute in 1942, the Atoms for Peace Award in 1967, the Oersted Medal from the American Association of Physics Teachers in 1982, the Four Freedoms Award from the Franklin and Eleanor Roosevelt Institute in 1985, and the Vannevar Bush Award from the National Science Foundation in 1985.

Although he lived and worked in various places in the United States and Europe, Rabi was always a New Yorker. He married Helen

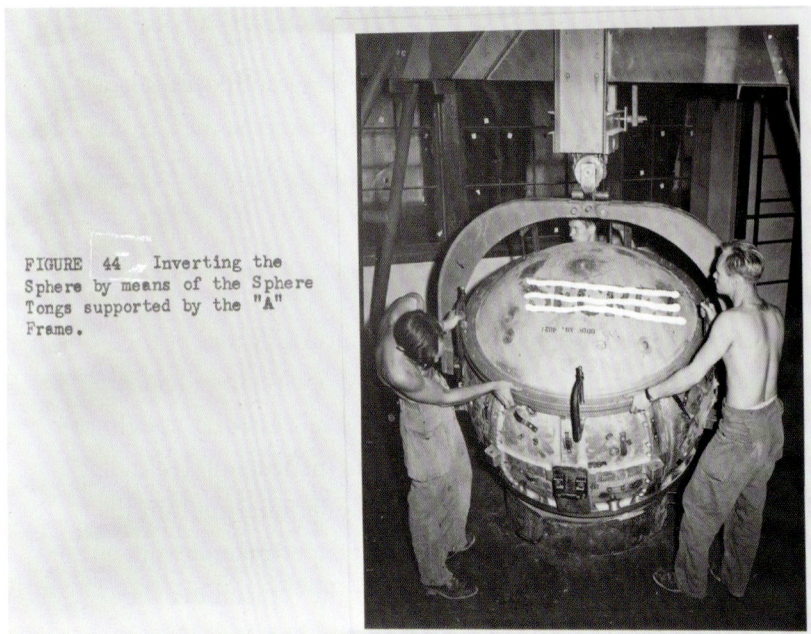

FIGURE 44 Inverting the Sphere by means of the Sphere Tongs supported by the "A" Frame.

Fat Man was the second of two atomic bombs ever used in combat. Thought to have been named after British Prime Minister Winston Churchill, the weapon weighed more than 10,000 pounds.

His influence in the world of physics led many to call him "the dean of world physics."

Newmark in 1926 and raised two daughters, Helen and Margaret. Rabi died in 1988, six months short of his 90th birthday. He was at home in New York.

Rabi believed he decided to become a scientist while growing up in Brooklyn. "My mother made me a scientist without ever intending to," he once said. "Every other Jewish mother in Brooklyn would ask her child after school: 'So? Did you learn anything today?' But not my mother. 'Izzy,' she would say, 'did you ask a good question today?' That difference — asking good questions — made me become a scientist."

After World War II ended, Rabi went on to research nuclear magnetic resonance, which lead to the MRI machine in use in the field of medicine.

ISIDOR RABI
1898–1988

LOS ALAMOS CONTRIBUTIONS

Helped change the laboratory's organizational concept from a military-led coalition to one led by civilian scientists. He served as a consultant on the Manhattan Project, acted as a personal adviser to J. Robert Oppenheimer, and was a frequent lifelong visitor to the lab.

NOBEL PRIZE: PHYSICS, 1944

Earned the Nobel Prize for Physics in 1944 for his work in developing a magnetic resonance method capable of observing atomic spectra. This method enables scientists to measure the magnetic properties of atoms, atomic nuclei, and molecules.

NORMAN RAMSEY

By Madeline Whitacre

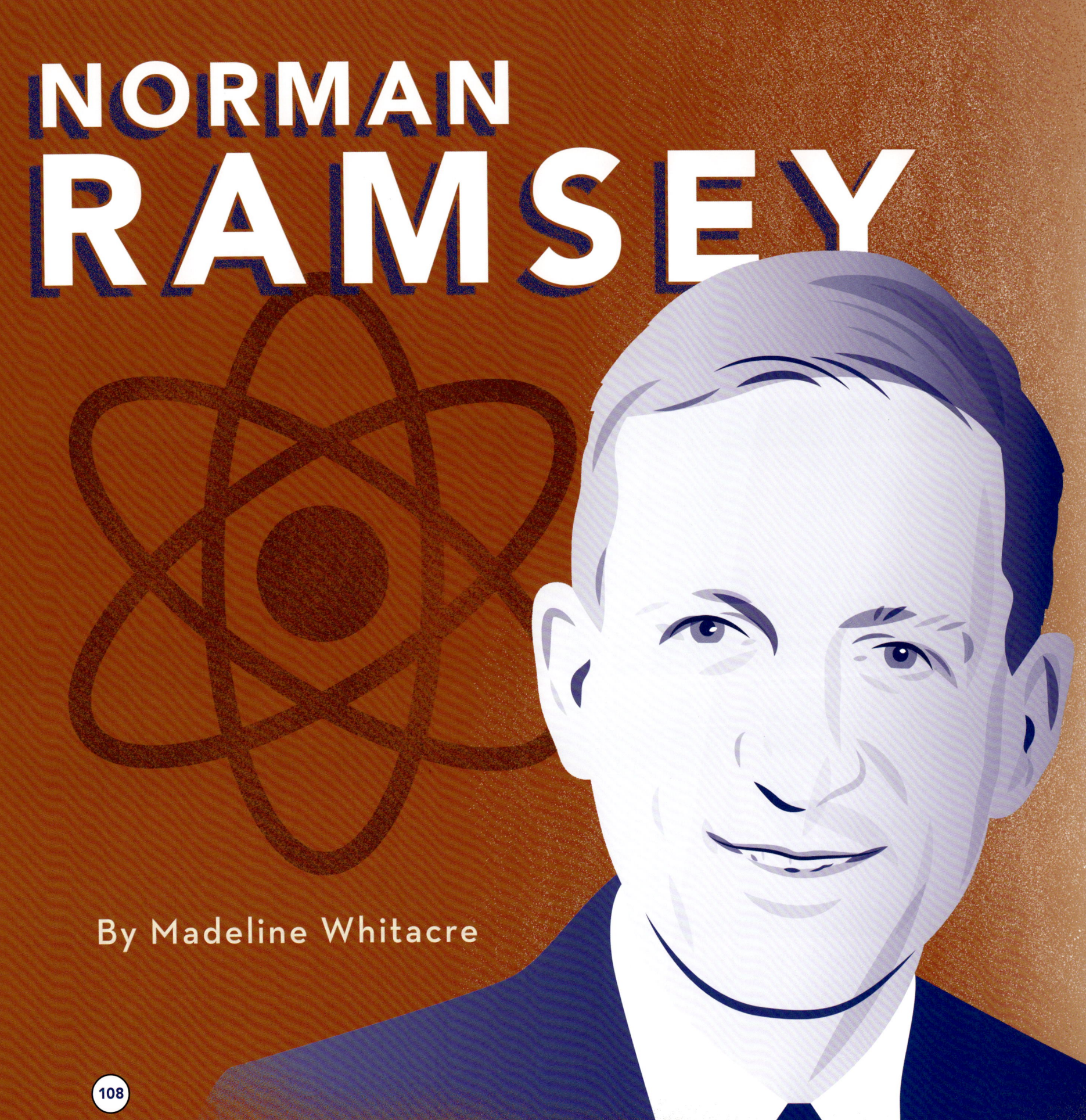

Norman Ramsey was known for snow skiing (even after having a knee replaced), longboard surfing, and ice sailing. But when given an employee personnel questionnaire at the lab in Los Alamos, he offered only one answer for his hobbies: physics. It was a pastime — and a career — that made an impact. In 1989, years after leaving Los Alamos where he worked on the world's first nuclear weapons, Ramsey won the Nobel Prize in Physics.

Early years

Ramsey was born in Washington, D.C., on August 27, 1915. His father was an Army officer and he moved often during his childhood. After graduating from high school at 15 years old, Ramsey enrolled in Columbia University in New York City, graduating with a degree in mathematics in 1935. Following his graduation, Ramsey earned a fellowship to attend the University of Cambridge in the United Kingdom, where he met and studied under exemplary physicists. It was at Cambridge that Ramsey became interested in molecular beams, on which he wrote his Ph.D. dissertation after his return to Columbia.

In 1940, Ramsey married Elinor Jameson. Shortly thereafter, the couple moved to Massachusetts and Ramsey joined the MIT Radiation Laboratory to work on radar development in support of the war effort as a consultant to the secretary of war's office. Ramsey was later recruited by physicist and Los Alamos Lab Director J. Robert Oppenheimer to join the Manhattan Project, which was the U.S.-led effort during World War II to develop the first atomic bombs. As a highly valued consultant, Ramsey remained on the payroll of the secretary of war office, but was on loan to the secret weapons laboratory in northern New Mexico. Ramsey and his family arrived in Los Alamos on September 27, 1943.

Los Alamos work

Until his departure in March 1945, Ramsey was a group leader in the Ordnance Division. Ramsey's group was responsible for preparing to use bombs as combat weapons. Ramsey first needed to determine which aircraft would be capable of carrying the atomic weapons that were being designed. At this time, the laboratory was still working on a plutonium gun-type weapon, dubbed Thin Man. Scientists ultimately abandoned the Thin Man-type weapon, and instead developed a plutonium implosion weapon called Fat Man. Efforts to develop a gun-type weapon shifted to using uranium in what would eventually become Little Boy — the first of two Los Alamos-created atomic bombs released on Japan.

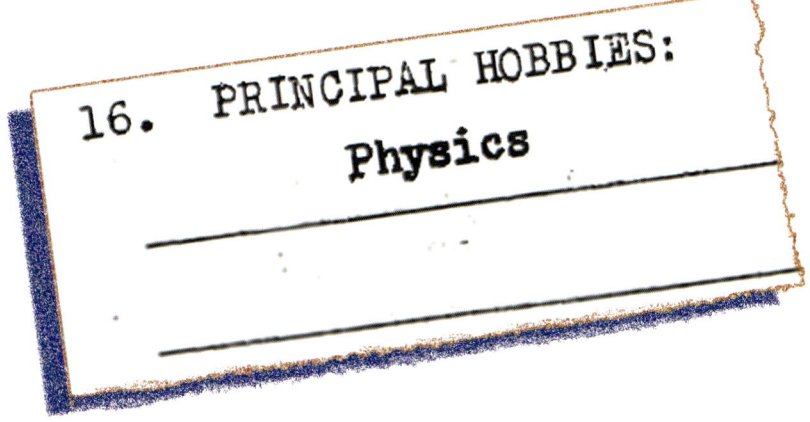

On his personnel questionnaire from the Los Alamos wartime lab, Norman Ramsey listed just physics as his hobby. He later went on to win the 1989 Nobel Prize in Physics.

gate into a tuned resonant cavity. Some atoms then move to a lower energy level resulting in a microwave signal at the resonance frequency of hydrogen. A quartz crystal oscillator then receives this signal and stays in tune with the hydrogen resonance frequency.

Ramsey's work impacts our world today. Because of their precision, atomic clocks are still used as the time standards for GPS satellites, and have been used to test the theory of relativity. The use of Ramsey's method has also led to the development of MRI scanners, which generate images of the organs in the body.

Ramsey died on November 4, 2011 at age 96. He was survived by his second wife, Ellie; four daughters, Margaret, Patricia, Winifred, and Janet; and two stepchildren, Marguerite and Gerard.

Scientists from the Manhattan Project reunited in 1983. Norman Ramsey is seated, bottom left. Other Nobel laureates pictured here include Emilio Segrè, Luis Alvarez, and Isidor Rabi.

Norman Ramsey, second row, far left, joins scientists and other staff from Los Alamos, as well as military personnel, for a briefing. The group was on Tinian island and meeting just prior to the U.S. strike on Hiroshima, Japan, where Little Boy was deployed. It was the first of two nuclear weapons to be used in combat. (Photo courtesy of Harold Agnew, future third Los Alamos lab director, who also attended the briefing.)

Norman Ramsey's badge photo was issued upon arrival in Los Alamos in 1943. His primary role was to help prepare the atomic weapons for combat delivery.

NORMAN RAMSEY, JR.

1915–2011

LOS ALAMOS CONTRIBUTIONS

Worked as the group leader in the Ordnance Division from 1943-1945; Deputy for Scientific and Technical Matters in Project A, 1945; prepared bombs for use in combat.

NOBEL PRIZE: PHYSICS, 1989

Developed a precise method to probe the structure of atoms and molecules, using the method to keep time precisely.

Nobel Laureates of Los Alamos
QUOTES

"Winning the prize wasn't half as exciting as doing the work itself."

-Maria Goeppert Mayer

"There are two possible outcomes: if the result confirms the hypothesis, then you've made a measurement. If the result is contrary to the hypothesis, then you've made a discovery."

-Enrico Fermi

(LANL photo by L.D.P King.)

"There are questions which illuminate, and there are those that destroy. I was always taught to ask the first kind."

-Isidor Rabi

"If we fight a war and win it with H-bombs, what history will remember is not the ideals we were fighting for, but the methods we used to accomplish them."

-Hans Bethe

"I would rather have questions that can't be answered than answers that can't be questioned."

-Richard Feynman

"An expert is a man who has made all the mistakes which can be made, in a narrow field."

-Niels Bohr

"You have to get a little untrapped from too much prior knowledge."

-Norman Ramsey

"I was strongly encouraged by a science teacher who took an interest in me and presented me with a key to the laboratory to allow me to work whenever I wanted."

-Frederick Reines

"The whole structure of science gradually grows, but only as it is built upon a firm foundation of past research."

-Owen Chamberlain

"I saw science as being in harmony with humanity."

- Sir Joseph Rotblat

"The constant questioning of our values and achievements is a challenge without which neither science nor society can remain healthy."

-Aage Bohr

"I remember the spring of 1941 to this day. I realized then that a nuclear bomb was not only possible — it was inevitable."

-Sir James Chadwick

"I see a tremendous amount of intricacy in the world and we have probably only begun to scratch at the surface of its intricacy."

-Roy Glauber

"In scientific matters there was a common language and one standard of values; in moral and political problems there were many."

-Emilio Segrè

"[M]y idea of science is now something that you do and learn, trying to learn new things."

-Edwin McMillan

"Whatever we were able to achieve in our later years had its origin in the experiences of our youth and in the hopes and wishes which were formed before and during our time as students.

-Felix Bloch

"[My father] advised me to sit every few months in my reading chair for an entire evening, close my eyes and try to think of new problems to solve. I took his advice very seriously and have been glad ever since that he did."

-Luis Alvarez

"It is highly improbable, a priori, to begin life on a cattle ranch and then appear in Stockholm to receive the Nobel Prize in Physics. But it is much less improbable to me when I reflect on the good fortune I have had in the ambience provided by my parents, my family, my teachers, colleagues and students."

-Val Fitch

FREDERICK REINES

By Mott Linn

Although his work on neutrinos — subatomic particles produced during nuclear fusion, which powers the sun, stars, and even nuclear reactors — led him to win the Nobel Prize in Physics, Frederick Reines as a young man was initially more interested in literature, history, and music. Indeed, Reines was a gifted baritone and excelled equally in chorus and solo performances.

"The first stirrings of interest in science that I remember occurred during a moment of boredom at religious school, when, looking out of the window at twilight through a hand curled to simulate a telescope, I noticed something peculiar about the light," Reines once recalled. "It was the phenomenon of diffraction. That began for me a fascination with light."

By the time he graduated high school, Reines had identified his pursuit with certainty, as exemplified by the goal he wrote in his high school yearbook: to become "a physicist extraordinaire."

Higher education

Born in Patterson, New Jersey, in 1918, Reines grew up around the New Jersey-New York border near the Hudson River. He attended the Stevens Institute of Technology in Hoboken, New Jersey, where he earned a bachelor's degree in engineering in 1939 and a master's degree in mathematical physics in 1941. He then attended New York University, where he completed a doctoral program in physics. Reines' doctoral research, which built upon the original fission calculations by physicists Niels Bohr (see Page 42) and John Wheeler, garnered the attention of Manhattan Project scientists who were working on the world's first atomic bombs.

Frederick Reines, behind the sign, is pictured here with other members of the team that went to the lab in Hanford, Washington, to prove that neutrinos exist. Their work was named Project Poltergeist because of the ghostly nature of the neutrino. Reines' collaborator, Clyde Cowan, is on the upper right.

"[He was] talented in both theory and experiment," Wheeler said of Reines, "a bear of a man given to thinking big about nearly impossible problems as he paced up and down in his oversized shoes."

Fifteen-year career at Los Alamos

Close to finishing a Ph.D. thesis on nuclear fission while attending New York University, Reines was recruited to a clandestine lab in Los Alamos, New Mexico, that was part of the Manhattan Project, the U.S. government's top-secret effort to create the first atomic bomb. Upon arriving at Los Alamos in 1944, Reines joined the project's Theoretical Division, which was led by Hans Bethe (1967 Nobel Prize for Physics, see Page 24).

Reines worked in a theoretical group led by Richard Feynman (see Page 70), where he and fellow scientists investigated the theory of diffusion and how it influenced the calculation of critical mass needed to create a sustained nuclear reaction. In June 1946, Reines became the leader of T-1, the Theory of Dragon Group. The group was so named because of the "tickling the dragon's tail" experiment, in which the dragon represented a machine capable of attaining criticality, or when a nuclear chain reaction is self-sustaining for short bursts of time.

Unlike many of the Manhattan Project scientists, Reines elected to remain in Los Alamos after the end of World War II in the fall of 1945. He shifted his focus to studying the effects of nuclear weapons and becoming an expert on blast waves and the radiation pollution caused by nuclear explosions. He participated in a number

Los Alamos staff were assigned varying levels of security that were strictly enforced.

Frederick Reines remained at the lab following World War II, participating in a number of nuclear tests, including Operation Crossroads, pictured here.

Ramsey, Norman
Arrived: 9/27/43

visit   115 B

with Parsons

wife Eleanor J. and baby

Terminated Oct. 3, 1945   Consultant

Arrival: Dec. 3, 1945
To See: Mr. Bradbury

Reines's employment card, which reveals his mobility in the laboratory's various technical areas after the war.

of nuclear tests, including Operations Crossroads, Operation Sandstone, and Operation Buster-Jangle. Moreover, he served as director for Operation Greenhouse, which tested the first boosted fission weapons with increased yield due to the addition of small quantities of fusion fuel. These weapons would subsequently lead to the development of thermonuclear weapons, more commonly known as H-bombs.

Chasing the elusive neutrino

As early as 1947 or so, while still at Los Alamos, Reines began to flirt with the idea of observing a neutrino, a revolutionary concept

that was little more than theory at the time. He asked to switch from testing nuclear weapons to this area of fundamental physics research. Although the concept of the neutrino had been around since the 1930s, no one had experimentally proven their existence — this is exactly what Reines proposed to accomplish. He spoke with Los Alamos physicist Clyde Cowan at length about his idea, and the two decided to collaborate.

Although the collaborators considered detecting neutrinos during an atomic blast, they decided that the explosion's brevity would make such detection much too difficult. They then reasoned that a nuclear reactor would be ideal, as it would be a much more controlled environment. In 1955, the men traveled to the Savannah River Site in South Carolina, where a year later they experienced their eureka moment. Reines and Cowan had detected neutrinos emitted by the reactor, recording the presence of these subatomic particles with the help of protons in two tanks of water.

In 1957, Cowan left Los Alamos to accept a professorship at George Washington University. Two years later, Reines too left Los Alamos to lead the physics department at the Case Institute of Technology in Cleveland, Ohio. There, Reines established a neutrino research team that would go on to be the first to detect neutrinos created by cosmic rays in the Earth's atmosphere.

Reflecting on his time in Los Alamos, Reines noted, "I worked in the company of perhaps the greatest collection of scientific talent the world has ever known."

Life after the Atomic City

After almost a decade at the Case Institute of Technology, Reines became the first dean of the School of Physical Sciences at the then-recently opened University of California at Irvine. He brought with him the majority of his colleagues from the neutrino research team.

In 1995, Reines received the Nobel Prize for his work on neutrinos. He was the only former Manhattan Project scientist to be awarded the prize for work conducted as a Los Alamos staff member. Cowan, who died in 1974, was not honored because Nobel Prizes are not awarded posthumously.

Among Reines' other awards were the National Medal of Science, the Franklin Medal (awarded by the Benjamin Franklin Institute Committee on Science and the Arts), the Bruno Rossi Prize, and the J. Robert Oppenheimer Memorial Prize. He also was named to the National Academy of Sciences, for which members are elected based on outstanding achievements.

"I worked in the company of perhaps the greatest collection of scientific talent the world has ever known."

Frederick Reines (foreground) and Clyde Cowan working together in the early 1950s. Their partnership led to the 1956 discovery of evidence that neutrinos exist. Cowan died before his contributions could be recognized with a Nobel Prize, which is not awarded posthumously.

Reines died on August 26, 1998, in Irvine. He was survived by his wife of 58 years, Sylvia; a son, Robert; and a daughter, Alisa.

In their 2009 biographical memoir, William Krop, Jonas Schultz, and Henry Sobel wrote, "Frederick Reines was a man of imposing physical stature, with an even more imposing appetite for physics and a passion for discovery. His energy, drive, and far-reaching vision carried him to the very heights of discovery but never quite satisfied his yearning for more. His philosophy could be described by a line from [British poet and playwright] Robert Browning he sometimes quoted, 'ah, but man's reach should exceed his grasp/Or what's a heaven for?'"

Federick Reines received the Nobel Prize in 1995 for his work on neutrinos. He was the only former Manhattan Project scientist to be awarded the prize for work conducted as a Los Alamos staff member.

FREDERICK REINES

1918–1998

LOS ALAMOS CONTRIBUTIONS

Worked as the Group Leader in the Theoretical Division; after World War II, took part in various Los Alamos nuclear tests and began preliminary work on proving the neutrino's existence.

NOBEL PRIZE: PHYSICS, 1995

Detection of the neutrino, which is is a subatomic particle similar to an electron, but has no electrical charge and is nearly massless and travels at near lightspeed.

SIR JOSEPH ROTBLAT

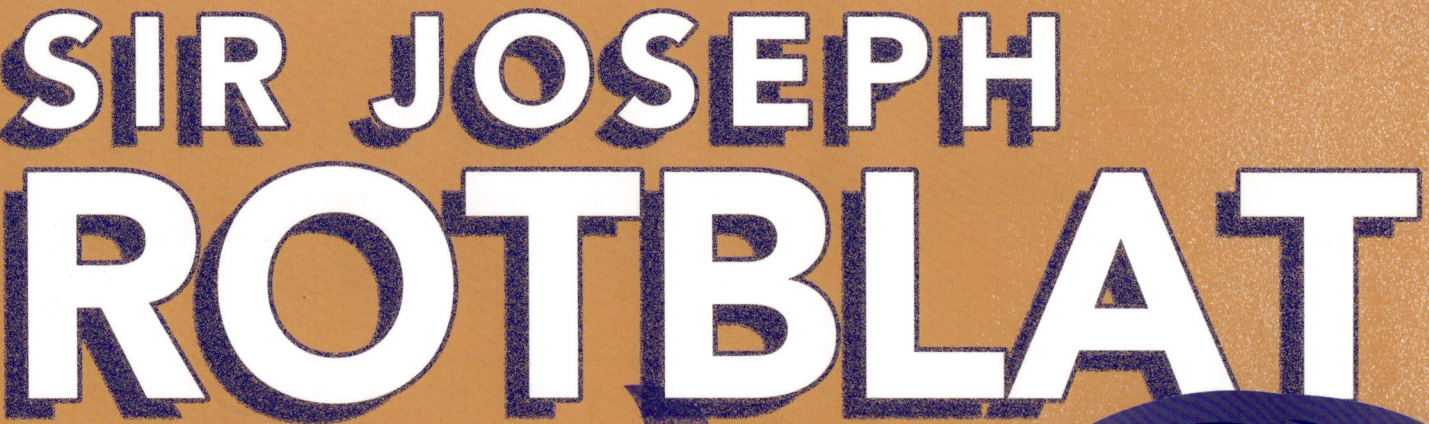

By Richard Moore

Unique among the 18 Nobel laureates with a wartime connection to Los Alamos National Laboratory, Joseph Rotblat's prize was not for physics or chemistry. As a physicist, Rotblat did make significant scientific discoveries in prewar Poland and also contributed to experimental work at Los Alamos. However, Rotblat's Nobel was the Peace Prize for his anti-nuclear weapons efforts, awarded in 1995. Rotblat said he had been "trying for 40 years to save the world, sometimes against the world's wishes."

Early years

Rotblat was born in Warsaw on November 4, 1908, in what was then Russian Poland. During World War I his father's business was ruined, and, as Jews, the family was subject to routine discrimination. Destitute, the young Rotblat "was bullied in food queues, felt the crump of artillery fire … and witnessed deaths through violence or illness," according to his biographer Andrew Brown.

After the war, Rotblat began work as an electrician's apprentice, but with ambitions to become a physicist, he later enrolled in the Free University of Warsaw. There, Ludwik Wertenstein, a humanitarian scientist who had worked with both Marie Curie (two-time Nobel Prize winner in physics and chemistry; first female recipient of the prize) and Ernest Rutherford ("the father of nuclear physics"), became his mentor and a major influence on his life. In 1935, Rotblat married Tola Gryn, a student of Polish language and literature. He earned his Ph.D. in physics in 1938 and served as assistant director of the Free University's Atomic Physics Institute.

```
Rotblat, Jozef                    Polish cit.
Arrived: 2/28/44

visitor
with Bacher P. O.
```

Joseph (who first went by Jozef) Rotblat's Los Alamos inprocessing card notes his Polish citizenship. Rotblat, who was Jewish, came to the United States via England. His desperate efforts to get his wife out of Poland were unsuccessful and she died in a Nazi concentration camp.

With few resources, Rotblat became an ingenious experimentalist. He confirmed the inelastic scattering of neutrons, and was the first to identify cobalt-60, now a significant industrial and medical radioisotope. Early in 1939 he discovered fission neutrons, independently of Frédéric Joliot-Curie's team, although delay in translating his paper meant the French research was published first.

Rotblat quickly realized that an explosive chain reaction was now a possibility. In April of that year, he moved to Liverpool, England to work on James Chadwick's new cyclotron. Chadwick (see Page 48) was a Nobel laureate and would become Rotblat's fellow Manhattan Project scientist. Lonely, and struggling with the language, Rotblat didn't share his concerns with Chadwick until November, by which time his boss was already involved in government inquiries into the feasibility of an atomic bomb.

Rotblat made a short visit to Warsaw in August 1939, but was unable to bring Tola back to England. She was recovering from

Joseph Rotblat joined the Manhattan Project's lab in Los Alamos, where scientists and other staff secretly worked to create atomic bombs to help end World War II.

appendicitis and would join him when she could travel. Tragically, he never saw her again. World War II intervened and Rotblat's desperate efforts to get her out of Poland through Denmark, Belgium, and Italy all failed. Tola was murdered by the Nazis in a concentration camp. Rotblat never remarried.

Los Alamos laboratory

In 1943, when Chadwick became head of the British atomic energy mission to the United States, he initially left Rotblat in charge in Liverpool. But Rotblat soon followed his boss to Los Alamos, arriving in early 1944. Rotblat stayed in Chadwick's house on Bathtub Row (nicknamed for the town's only homes with bathtubs, which were given to the most senior laboratory staff) until Chadwick's family arrived a month later, and Rotblat moved into bachelor quarters.

Rotblat's working life at Los Alamos was spent in nuclear physicist Robert Bacher's Physics Division, where Rotblat designed and conducted experiments, such as measuring fission and capture cross-sections and neutron energies, and observing and understanding prompt fission gamma rays.

Joseph Rotblat's focus at Los Alamos was fission, or when atoms are split, and the relation to nuclear energy. Rotblat used an apparatus, right, to count and measure the energy of fission gamma rays.

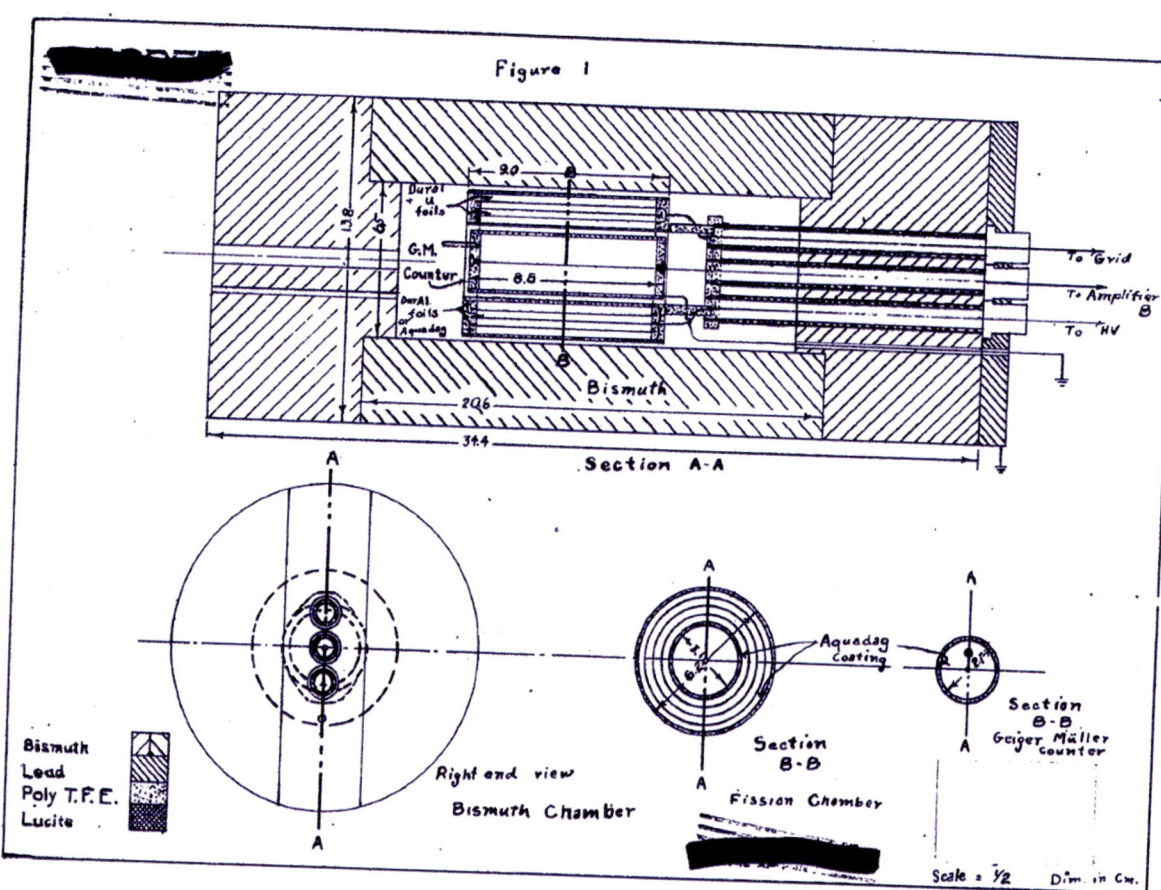

Later, he worked on the Los Alamos cyclotron, observing fast-neutron effects on fission products, and developed detection equipment for physicist Otto Frisch's "dragon-tickling" nuclear criticality experiments, which involved bringing two or more pieces of fissile material together to form a critical mass to start an atomic chain reaction.

Rotblat was concerned from early on about the relevance of his work in the fight against Nazi Germany. Soon after his arrival, he was horrified to hear Manhattan Project leader General Leslie Groves talk about the real purpose of the bomb — in Groves's view, "to subdue the Russians." In November 1944, when Chadwick told Rotblat about intelligence that the Germans' efforts to create their own atomic bomb had failed, Rotblat's mind was made up. He asked to leave Los Alamos and the project. Although his request was granted, it aroused unsubstantiated suspicions about his loyalty. As he was leaving the country, Rotblat's trunk full of papers mysteriously disappeared on the train between Washington, D.C., and New York. Rotblat was banned from the United States for many years.

Post-World War II

Back in Liverpool, Rotblat at first remained silent about his misgivings about the atomic bomb. He remained a member of the British atomic energy team and worked, for example, on plans for the post-war Atomic Energy Research Establishment in Harwell, England. Rotblat had also become a British citizen, which allowed him to help family members join him in England.

By the end of 1945, Rotblat resolved that having contributed to what he saw as science's misuse at Los Alamos, he would devote the rest of his life to campaigning against nuclear weapons and against war. In 1950 he accepted a chair position at St. Bartholomew's teaching hospital in London, where he worked in nuclear medicine until his retirement.

In 1957, Rotblat organized the first in a series of international conferences that would become his most important legacy: the Pugwash Conferences on Science and World Affairs, which continue today. Cyrus Eaton, a millionaire industrialist, offered funding for the first meeting in his tiny hometown of Pugwash,

As Manhattan Project veteran Otto Frisch put it, Pugwash could roam "where diplomats fear to tread." Rotblat was the organization's guiding light for 50 years.

Nova Scotia. Manhattan Project veterans, including Mark Oliphant, Eugene Rabinowitch, Leo Szilard, and Victor Weisskopf were among the guests.

From these small beginnings, Pugwash has grown into a significant global movement that brings together scholars and public figures to seek solutions to global security threats without nuclear weapons or other weapons of mass destruction. Pugwash was also an important back channel in Cold War arms-control negotiations including, notably, the Partial Test Ban Treaty (1963) and the Anti-Ballistic Missile Treaty (1972). As Manhattan Project veteran Otto Frisch put it, Pugwash could roam "where diplomats fear to tread." Rotblat was the organization's guiding light for 50 years.

Housing and community buildings, such as a school, hospital, and store, were quickly constructed along with infrastructure for the lab itself.

In 1995, Rotblat and the Pugwash movement were jointly awarded the Nobel Peace Prize for their influential work for peace in the nuclear age. That same year, Rotblat was named a fellow of the Royal Society, the United Kingdom's National Academy of Sciences, and was knighted by Queen Elizabeth II in 1998.

Rotblat continued writing, traveling, and lecturing until shortly before his death on August 31, 2005 at age 96, leaving behind no immediate family.

Joseph Rotblat's identification badge photo was taken shortly after his arrival to New Mexico. He came in 1944, but left less than a year later due to his misgivings about developing the atomic bomb. After World War II, Rotblat believed he had contributed to the misuse of science and devoted the rest of his life to campaigning against nuclear weapons and war.

SIR JOSEPH ROTBLAT

1908–2005

LOS ALAMOS CONTRIBUTIONS

Experimental work in support of fundamental nuclear measurements.

NOBEL PRIZE: PEACE, 1995

Developed the nuclear magnetic resonance technique for studying the composition of materials.

EMILIO SEGRÈ

By Megan Jochem

When most children his age were learning to read, count, and tell time, Emilio Segrè was working on physics. "[W]hen I was six years old—because it happens to be dated—I had a school book, a notebook, in which it says, 'Physics experiments by Emilio Segrè,'" he recalled during an oral history interview conducted for the American Institute of Physics in May 1964.

"I wrote [down] all these little experiments that I was doing, which I read in books more or less for children but that I liked to do myself. I had the spectrum of the sun, the atmospheric pressure — all my little experiments. So, that's very, very early."

Colleagues described an adult Segrè as a complicated man, one who had high standards and expected others to measure up. On the outside, he was proud, aloof, and somewhat intimidating. But on the inside, Segrè was welcoming and generous when it came to supporting younger physicists, always ready with helpful advice and guidance. He possessed a European flair about him, capable of speaking several languages and quoting from the classics, such as Dante, Victor Hugo, and Friedrich Schiller. When not in his lab, Segrè took to the outdoors, mountaineering, fly fishing, and collecting wild mushrooms on rigorous hikes.

Meeting Enrico Fermi

Upon graduating from high school in 1922, Segrè attended the University of Rome La Sapienza, initially as an engineering student. While at the university, he attended mathematics seminars. Admitting he rarely understood the topics, Segrè still found the talks fascinating. It was at one of these seminars that Segrè encountered Enrico Fermi, who offered the clarity Segrè longed for and perhaps provided a hint of the future both men would have together decades later in the United States.

I remember I heard Dau [Russian theoretical physicist Lev Landau, whose nickname was "Dau"] speaking of number

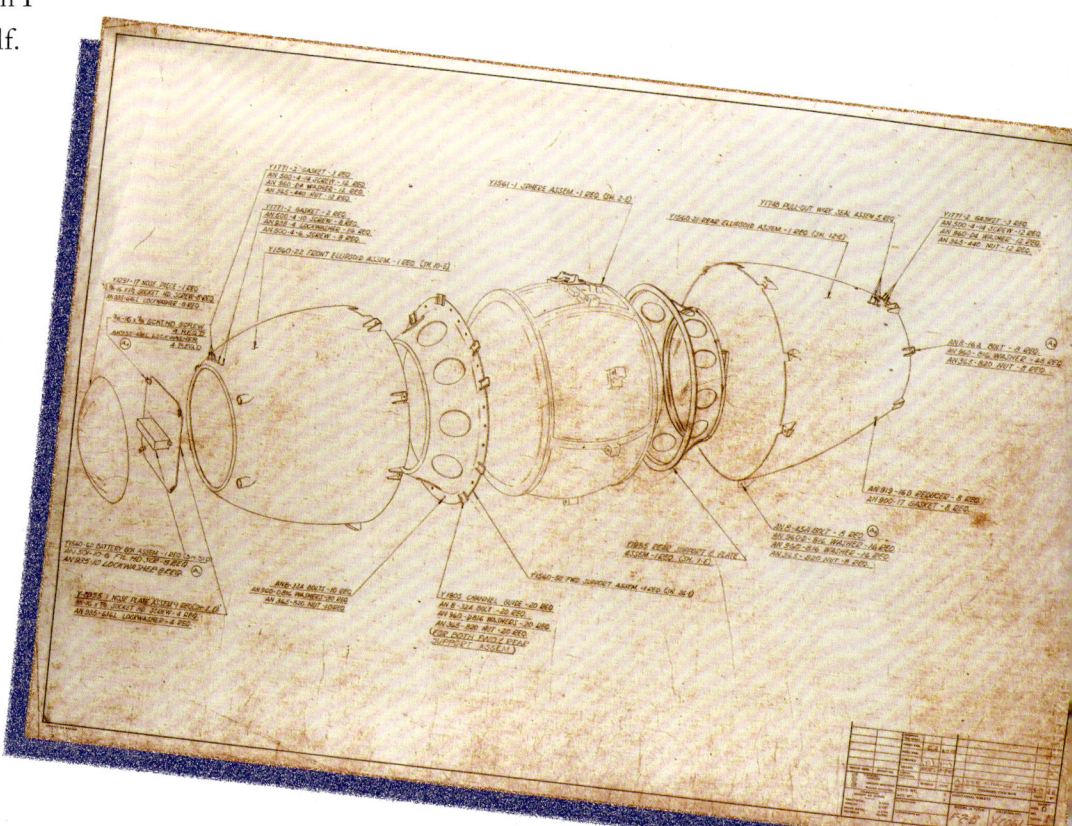

Emilio Segrè's work led to the decision to forgo the Thin Man weapon, which would not be successful, and to continue developing a plutonium, implosion-type bomb known as Fat Man, shown here, as well as a uranium gun-type bomb known as Little Boy.

theory; I mean, it was fantastic," recalled Segrè in an oral history interview. "And I heard all kinds of other things. But then once I heard Fermi speaking about quantum theory ... it was absolutely plain that here was a man who knew what he was talking about, beyond normal. It was absolutely plain.

In 1927, Segrè met American physicist Franco Rasetti, who formally introduced him to Fermi. Segrè switched from engineering to physics, subsequently becoming Fermi's assistant. Consequently, Segrè found himself doing a lot of writing, so much so that his friends predicted that such writings would constitute the bulk of his legacy instead of the science itself.

> "It is clear that since you will never get the Nobel Prize for Physics, you are preparing yourself for the Literature Prize," Segrè was jokingly told.

> Despite such teasing, Segrè proved himself a gifted physicist, earning his Ph.D. in just one year, graduating in 1928 — and later earning a Nobel Prize in Physics in 1958.

Military service and early discoveries

Upon graduating from the University of Rome in 1928, Segrè served two years in the Italian Army as a second lieutenant in an anti-aircraft artillery unit. Upon his discharge in early 1930, Segrè returned to academia, initially working as an assistant to various professors in Italy, Germany, and Holland.

From 1936 to 1938, Segrè served as the director of the physics laboratory at the University of Palermo in Italy. It was at this laboratory that Segrè discovered element 43, or technetium, the first artificially produced element. Today, an isotope (atoms of the same element that have the same number of protons and electrons but a different number of neutrons) of technetium, technetium-99m, is used in countless medical diagnostic procedures.

To study technetium's isotopes, Segrè made frequent trips to the University of California, Berkeley, because the isotopes were so short-lived they would not survive a trip from California back to Segrè's lab in Palermo. As World War II ravaged Europe, Segrè, who was Jewish, elected to remain in the United States, where his wife and son soon joined him.

Emilio Segrè, far left, and his Radioactivity Group at Los Alamos during World War II. Standing behind Segrè is future Nobel laureate Owen Chamberlain (see Page 58).

At first a visitor to Berkeley's Radiation Laboratory, Segrè subsequently became a research associate and, at the beginning of World War II, served as a lecturer in Berkeley's physics department. During this time, Segrè and his team discovered element 85, known as astatine. Today, an isotope of astatine, astatine-211, is used to treat cancer.

In the early 1940s, Segrè met J. Robert Oppenheimer, who at the time was a professor of theoretical physics. In 1943, Oppenheimer convinced Segrè to join him at Los Alamos as part of the Manhattan Project, the U.S. government's top-secret effort to build the first atomic bombs to help end World War II.

Work at Los Alamos

Upon arriving at Los Alamos in 1943, Segrè worked as the head of the Radioactivity Group, which was under the Experimental Physics Division.

Initially, Oppenheimer and colleagues worked to create a gun-type nuclear weapon called Thin Man. However, Segrè and his team eventually discovered issues in developing this type of atomic weapon, including the existence of plutonium-239. This is an isotope of plutonium and caused higher rates of spontaneous fission, which meant that this isotope was too unpredictable for use in a weapon. Segrè and his team thought plutonium-239 would evolve into plutonium-240, which was another type of spontaneous fission plutonium. This led to the decision to forgo Thin Man, which would not be successful, and to continue developing a plutonium, implosion-type bomb known as Fat Man, as well as a uranium gun-type bomb known as Little Boy.

By the summer of 1945, Segrè and his colleagues were preparing for the Trinity test — the first-ever detonation of an atomic device — that would take place on July 16 in the New Mexico desert. Segrè and several other colleagues were tasked with measuring gamma rays during the explosion of the device, which was called The Gadget. A weaponized version of The Gadget, called Fat Man, was used in combat just weeks later. Meanwhile, although a full-scale nuclear explosive test was not conducted for the Little Boy atomic bomb, every component of the weapon was rigorously tested at the Los Alamos lab. As such, the scientists were mathematically certain it would be successful in combat.

Of the Trinity device detonation, Segrè recalled: "In a fraction of a second, at our distance, one received enough light to produce a sunburn. I was near Fermi at the time of the explosion, but I do not remember what we said, if anything. I believe that for a moment I thought the explosion might set fire to the atmosphere and thus finish the Earth, even though I knew that this was not possible."

Weeks later, on August 6 and 9, respectively, Little Boy and Fat Man were released above Japan. World War II officially ended on September 2, 1945.

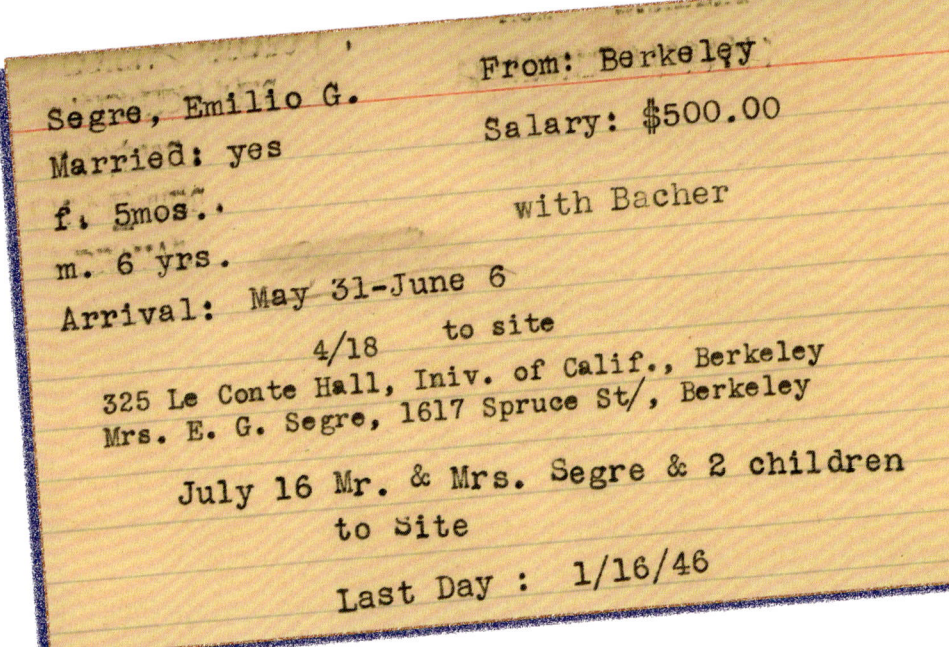

Famed physicist J. Robert Oppenheimer convinced Emilio Segrè to join him at Los Alamos as part of the Manhattan Project, the U.S. government's top-secret effort to build the first atomic bombs to help end World War II.

In limited amounts. Like other forms of antimatter, they may be useful in propulsion by energizing a propellant or heating a solid-fuel core.

Once Segrè and his team discovered that antiprotons were making appearances in the Bevatron, they began working on publishing a paper with their findings. In 1959, four years after the discovery of the antiproton, Segrè and Owen Chamberlain received news that they had won the Nobel Prize in Physics.

After traveling around the world, giving lectures, and teaching, Segrè eventually retired and settled in Lafayette, California. After the death of his first wife, Segrè remarried to Rosa Mines in 1972. Segrè died in April 1989. He was survived by Rosa and his three children, Amelia, Claudio, and Fausta.

Physicists Philip Morrison (left) and Emilio Segrè share a moment of levity. Both worked at the secret lab in Los Alamos during World War II, helping to create the first-ever nuclear weapons.

In remembering the Trinity test, Segrè captured the essence of that scientific achievement and its implications for the world: "Even though the purpose was grim and terrifying," he noted, "it was one of the greatest physics experiments of all time."

Antiproton discovery leads to Nobel Prize

In 1946, Segrè left Los Alamos and returned to Berkeley to continue his work on spontaneous fission and astatine. During his later years, he discovered the antiproton, for which he would be awarded science's most prestigious prize. Segrè applied the theory of the positron, which had been studied by Paul Dirac in the late 1920s and officially discovered in 1932 by C. D. Anderson.

Segrè believed that with every particle there existed an antiparticle. The experiments took place in 1955, when Segrè and his team used a Bevatron machine to help in the discovery of the antiproton. A particle accelerator, which produces beams of charged particles for research purposes, the Bevatron helped prove the existence of the antiproton because it contained enough energy to produce such subatomic particles. Antiprotons can only be produced in very large facilities and

During his later years, Segrè discovered the antiproton, for which he would be awarded science's most-prestigious prize.

Emilio Segrè worked as the head of the Radioactivity Group at the lab in Los Alamos. Among other work, Segrè helped measure gamma rays during the explosion of The Gadget, which was a plutonium, implosion-type device tested in the New Mexico desert. Its successful detonation marked the dawn of the Atomic Age.

EMILIO SEGRÈ

1905–1989

LOS ALAMOS CONTRIBUTIONS

Observed and studied The Gadget detonated during the Trinity test, which marked the dawn of the Atomic Age. Discovered that plutonium-240 made the Thin Man plutonium-weapon concept not viable, ultimately leading to the creation of the implosion plutonium device called Fat Man.

NOBEL PRIZE: PHYSICS, 1959

Discovered the antiproton, which can only be produced in very large facilities and in limited amounts, like other forms of antimatter, may be useful in propulsion, to produce direct thrust, by energizing a propellant or heating a solid fuel core.

Works Cited

Luis Alvarez

"Luis Alvarez." Atomic Heritage Foundation. Accessed February 23, 2021. https://www.atomicheritage.org/profile/luis-alvarez.

"The Nobel Prize in Physics, 1968." NobelPrize.org. Nobel Prize Outreach AB, 2021. Accessed February 23, 2021. https://www.nobelprize.org/prizes/physics/1968/summary/.

Alvarez, L. W. Alvarez: *Adventures of a Physicist*. New York: Basic Books, 1987.

Hoddeson, Lillian, Paul W. Henriksen, Roger A. Meade, and Catherine Westfall. *Critical Assembly: A Technical History of Los Alamos During the Oppenheimer Years, 1943–1945*. Cambridge: Cambridge University Press, 1993.

Maugh, T. H., II. "Physicist Luis Alvarez, 77, Nobel Prize Winner, Dies." *L.A. Times*. September 02, 1988. https://www.latimes.com/archives/la-xpm-1988-09-02-mn-2278-story.html.

Rhodes, Richard. *The Making of the Atomic Bomb*. New York: Simon & Schuster Inc, 1986.

Hans Bethe

"Hans Bethe." Atomic Heritage Foundation, 2019. Accessed February 23, 2021. https://www.atomicheritage.org/profile/hans-bethe.

"Web of Stories – Life Stories of Remarkable People." In Hans Bethe – Beginning physics at Frankfurt University, edited by Hans Bethe. London: Web of Stories Ltd, Science Navigation Group, 2017. Web series. https://www.youtube.com/watch?v=LvgLyzTEmJk&list=PLVV0r6CmEsFyUDS-roBQVEcbnNud7I9xom&t=0s.

"The Nobel Prize in Physics, 1967." NobelPrize.org, Nobel Prize Outreach AB, 2021. Accessed February 20, 2021. https://www.nobelprize.org/prizes/physics/1967/summary/.

Bernstein, J. Hans Bethe, *Prophet of Energy*. New York: Basic Books, 1980.

Broad, William J. "Hans Bethe, Prober of Sunlight and Atomic Energy, Dies at 98." *The New York Times* (New York), March 8, 2005. https://www.nytimes.com/2005/03/08/science/hans-bethe-prober-of-sunlight-and-atomic-energy-dies-at-98.html.

Broad, William J. "He Lit Nuclear Fire; Now He Would Douse It." *The New York Times* (New York), June 17, 1997. https://www.nytimes.com/1997/06/17/science/he-lit-nuclear-fire-now-he-would-douse-it.html.

Hoddeson, Lillian, Paul W. Henriksen, Roger A. Meade, and Catherine Westfall. *Critical Assembly: A Technical History of Los Alamos During the Oppenheimer Years, 1943-1945*. Cambridge: Cambridge University Press, 2004.

Rhodes, Richard. *The Making of the Atomic Bomb*. New York: Simon & Schuster Inc., 1986.

Felix Bloch

"Felix Bloch Facts." NobelPrize.org, Nobel Prize Outreach AB, 2021. Accessed February 20, 2021. https://www.nobelprize.org/prizes/physics/1952/bloch/facts/.

Bloch, Felix. "Nuclear Magnetism." *American Scientist* vol. 43, no. 1 (1955): 48-62.

Bloch, Felix. "Oral History Interview" by Charles Weiner. Stanford: University, California. August 15, 1968.

———. "Oral History Interview" By Thomas S. Kuhn. May 14, 1964.

Hartwig, Daniel. "Guide to the Felix Bloch Papers." Stanford University: Online Archive of California. Accessed February 20, 2021.

Hofstadter, Robert. "Felix Bloch." In National Academies of Sciences, Biographical Memoirs: Volume 64 Washington, DC: The National Academies Press, 1994.

Hofstadter, Robert, et al. "Memorial Resolution: Felix Bloch (1905-1983)." *Physics Today* vol. 37, no. 6 (1984): 81.

Kumar, Anil. "Felix Bloch (1905–1983)." *Resonance* vol. 20, no. 47 (2015): 956–58.

von Hippel, Arthur Robert. *Life in Times of Turbulent Transitions: The Autobiography of Arthur Robert von Hippel.* Princeton Junction: Stone Age Press, 1988. http://pdf.oac.cdlib.org/pdf/stanford/uarc/sc0303.pdf

Aage Bohr

"Aage N. Bohr – Biographical." NobelPrize.org, Nobel Prize Outreach AB, 2021. Accessed December 7, 2020. https://www.nobelprize.org/prizes/physics/1975/bohr/biographical/.

"Aage Bohr Biography." American Institute of Physics (AIP) Publishing. Accessed January 8, 2021. https://history.aip.org/phn/11707003.html.

"Ernest Rutherford Medal and Prize Recipients." Institute of Physics, 2021. https://www.iop.org/about/awards/silver-subject-medals/ernest-rutherford-medal-and-prize-recipients.

"Facts on the Nobel Prize in Physics." NobelPrize.org, Nobel Prize Outreach AB, 2021. Accessed February 24, 2021. https://www.nobelprize.org/prizes/facts/facts-on-the-nobel-prize-in-physics.

"Press Release." NobelPrize.org, Nobel Prize Outreach AB, 2021. Accessed December 6, 2020, https://www.nobelprize.org/prizes/physics/1975/press-release/.

Bacher, Robert. "Oral History Interview by Lillian Hoddeson on July 30, 1984 for the Book *Critical Assembly*." July 30, 1984.

Blaedel, Niels. *Harmony and Unity: The Life of Niels Bohr.* Scientific Revolutionaries: A Biographical Series. Madison, Wisconsin: Science Tech Publishers, 1988.

Bohr, Aage. "The War Years and the Prospects Raised by the Atomic Weapons." In *Niels Bohr: His Life and Work as Seen by His Friends and Colleagues*, edited by Stefan Rozental. New York: Interscience Publishers, John Wiley & Sons, Inc., 1967. 191-214.

———. Rotational States of Atomic Nuclei. Copenhagen: Ejnar Munksgaards Forlag, 1954.

Chang, Kenneth. "Aage Bohr, Physicist's Son Who Won Nobel, Dies at 87." *The New York Times* (New York), September 10, 2009. https://www.nytimes.com/2009/09/11/world/europe/11bohr.html

Critchfield, Charles. "The Robert Oppenheimer I Knew." In *Behind Tall Fences: Stories and Experiences About Los Alamos at Its Beginning*. Los Alamos, New Mexico: Los Alamos Historical Society, 1996.

Fermi, Laura. *Atoms in the Family: My Life with Enrico Fermi.* Chicago: The University of Chicago Press, 1961.

Frisch, Otto. *What Little I Remember.* Cambridge: Cambridge University Press, 1991.

Gowing, Margaret. "Niels Bohr and Nuclear Weapons." In Niels Bohr: A Centenary Volume, edited by A. P. French and P. J. Kennedy. Cambridge, Massachusetts: Harvard University Press, 1985. 266-77.

Hawkins, David, Edith C. Truslow, and Ralph Carlisle Smith. *Project Y: The Los Alamos Story.* The History of Modern Physics, 1800-1950. Vol. II, Los Angeles and San Francisco: Tomash Publishers, 1983.

Hirschfelder, Joseph. "The Scientific and Technological Miracle at Los Alamos." In *Reminiscences of Los Alamos, 1943-1945*, edited by Herbert Broida, Joseph Hirschfelder, Lawrence Badash. Studies in the History of Modern Science. Dordrecht, Holland: D. Reidel Publishing Company, 1980. 67-88.

Hoddeson, Lillian, Paul W. Henriksen, Roger A. Meade, and Catherine Westfall. *Critical Assembly: A Technical History of Los Alamos During the Oppenheimer Years, 1943-1945*. Cambridge, United Kingdom: Cambridge University Press, 1998.

Los Alamos Scientific Laboratory. "Among Our Guests." The Atom vol. 12, no. 5 (September-October 1975): 25.

Manhattan Engineer District. "Book VIII Los Alamos Project (Y), Volume 2 - Technical." In *Manhattan District History, Circa 1944-1946*. https://www.osti.gov/includes/opennet/includes/MED_scans/Book%20VIII%20-%20Volume%202%20-%20Technical.pdf

Marshak, Robert E. "Atoms for Peace Awards." *Science*, no. 164 (June 27, 1969): 1496-98.

Oppenheimer, J. Robert. "Niels Bohr and Atomic Weapons." *The New York Review* (New York), December 17, 1964, 6-8.

Richards, Hugh. "The Making of the Bomb: A Personal Perspective." In *Behind Tall Fences: Stories and Experiences About Los Alamos at Its Beginning*. Los Alamos, New Mexico: Los Alamos Historical Society, 1996. 116-34.

Solutions, Mosaic Architectural. Architectural Survey of Fuller Lodge Historic District. Los Alamos, New Mexico, 2008.

Szasz, Ferenc. *British Scientists and the Manhattan Project: The Los Alamos Years.* New York: St. Martin's Press, 1992.

Niels Bohr

"Nomination Archive: Christian Bohr." NobelPrize.org, Nobel Prize Outreach AB, 2021. https://www.nobelprize.org/

nomination/archive/show_people.php?id=1116

"The Nobel Prize in Physics 1922." NobelPrize.org, Nobel Prize Outreach AB, 2021. https://www.nobelprize.org/prizes/physics/1922/summary/.

Aaserud, Finn. "The Scientist and the Statesmen: Niels Bohr's Political Crusade During World War II." *Historical Studies in the Physical and Biological Sciences* vol. 30, no. 1 (1999).

Bernstein, Jeremy. "What Did Heisenberg Tell Bohr About the Bomb?" *Scientific American* 272, no. 5 (1995): 92-97.

———. Modulated Neutron Source.

Bohr, Aage. "The War Years and Atomic Weapons." In *Niels Bohr: His Life and Work as Seen by His Friends and Colleagues* edited by Stefan Rozental, 214. Amsterdam: North-Holland Publishing Co., 1967.

Bohr, Niels. "Open Letter to the United Nations." In *Niels Bohr: A Centenary Volume* edited by A.P. and P.J. Kennedy French. Cambridge: Harvard University Press, 1985. 295.

Cassidy, David C. "A Historical Perspective on Copenhagen." *Physics Today* 53, no. 7 (2000): 28-32.

Heisenberg, Werner. *Encounters with Einstein and Other Essays on People, Places, and Particles.* Princeton: Princeton University Press, 1983.

Hoddeson, Lillian, Paul W. Henriksen, Roger A. Meade, and Catherine Westfall. *Critical Assembly: A Technical History of Los Alamos During the Oppenheimer Years, 1943-1945.* Cambridge, United Kingdom: Cambridge University Press, 1998.

Moore, Ruth. *Niels Bohr: The Man, His Science, and the World They Changed.* New York: Alfred A. Knopf, 1966.

Nielson, Rud. "Memories of Niels Bohr." *Physics Today* 16, no. 10 (1963): 22-30.

Oppenheimer, J.R. "J.R. Oppenheimer to All Group and Division Leaders." Los Alamos: LANL, National Security Research Center, December 28, 1943.

Pais, Abraham. *Niels Bohr's Times, in Physics, Philosophy, and Polity.* Oxford: Clarendon Press, 1991.

Pais, Abraham. "Reminiscences from the Post-War Years." In *Niels Bohr: His Life and Work as Seen by His Friends and Colleagues* edited by Stefan Rozental. Amsterdam: North-Holland Publishing Co., 1967: 215.

Peierls, Rudolf. "Rutherford and Bohr." *Notes and Records of the Royal Society of London* vol. 42, no. 2 (1988): 229-30.

Rosenfeld, Léon and Erik Rüdinger. "The Decisive Years." In *Niels Bohr: His Life and Work as Seen by His Friends and Colleagues*, edited by Stefan Rozental. Amsterdam: North-Holland Publishing Co., 1967.

Rozental, Stefan, ed. *Niels Bohr: His Life and Work as Seen by His Friends and Colleagues* Amsterdam: North-Holland Publishing Co., 1967.

Wilson, Robert. "Niels Bohr and the Young Scientists." *Bulletin of the Atomic Scientists* vol. 41, no. 8 (1985): 24-6.

James Chadwick

Brown, Andrew. *The Neutron and the Bomb: A Biography of Sir James Chadwick.* Oxford: Oxford University Press, 1997.

Chadwick, James. "The Existence of a Neutron." Proceedings of the Royal Society of London. Series A, Containing Papers of a Mathematical and Physical Character vol. 136, no. 830 (1932): 692-708.

———. Report to the British Embassy in Washington. UK National Archives CAB 126/250 (July 16, 1945).

———. "Existence of a Neutron." Proceedings of the Royal Society A vol. 136, no. 830 (1932): 692-708.

———. "Possible Existence of a Neutron." *Nature* vol. 27 (1932): 312.

Chown, Marcus. "Norman Ramsey: Physicist Whose Work Led to the Atomic Clock and the Mri Scanner." *Independent* (London), 2011. https://www.independent.co.uk/news/obituaries/norman-ramsey-physicist-whose-work-led-atomic-clock-and-mri-scanner-6262579.html.

Feather, Harrie Massey and Norman. "Sir James Chadwick," Biographical Memoirs of Fellows of the Royal Society. 1976.

Gowing, Margaret. *Britain and Atomic Energy 1939-45*. New York: Macmillan, 1964.

Owen Chamberlain

Chamberlain, Owen. *The Early Antiproton Work*. 1959. Nobel Prize lecture.

Chamberlain, O., E. Segre, C.E. Wiegand, and T. Ypsilantis. "Observation of Antiprotons" *Physics Review*, no. 100 (1955): 947.

Dirac, P.A.M. "Proceedings of the Royal Society of London." 1930/1931.

———. "Proceedings of the Royal Society of London." 1930, 1931.

Feynman, Richard P. *Surely You're Joking, Mr. Feynman! (Adventures of a Curious Character)*. New York: W. W. Norton & Company, 1985.

Steiner, Herbert. *Owen Chamberlain, 1920-2006: A Biographical Memoir*. Washington, D.C.: National Academy of Sciences, 2010.

Enrico Fermi

"Enrico Fermi – Biographical." NobelPrize.org, Nobel Outreach AB, 2021, Accessed February 5, 2021. https://www.nobelprize.org/prizes/physics/1938/fermi/biographical/.

"Laura Fermi." Atomic Heritage Foundation, accessed February 3, 2021, https://www.atomicheritage.org/profile/laura-fermi.

"To Fermi ~ with Love." Voices of the Manhattan Project, accessed February 3, 2021, https://www.manhattanprojectvoices.org/oral-histories/fermi-love-part-1.

Bernstein, J. *Hans Bethe, Prophet of Energy*. New York: Basic Books, 1980.

Fermi, Enrico, My Observations During the Explosion at Trinity on July 6, 1945. Personal Communication.

———. Md-H-1 Form for Enrico Fermi. Folder 3 Box 14, (June 12, 1945).

Fermi, Laura. *Atoms in the Family: My Life with Enrico Fermi*. Chicago: University of Chicago Press, 1954.

Hoddeson, Lillian, Paul W. Henriksen, Roger A. Meade, and Catherine Westfall. *Critical Assembly: A Technical History of Los Alamos During the Oppenheimer Years, 1943-1945*. Cambridge: Cambridge University Press, 1993.

Rhodes, Richard. *The Making of the Atomic Bomb*. New York: Simon & Schuster Inc, 1986.

Segre, Emilio. *Enrico Fermi – Physicist*. Chicago: University of Chicago Press, 1970.

Richard Feynman

"The Nobel Prize in Physics, 1967." NobelPrize.org, Nobel Outreach AB, 2021. Accessed February 20, 2021. https://www.nobelprize.org/prizes/physics/1967/summary/.

Feynman, Richard P. *Surely You're Joking, Mr. Feynman! (Adventures of a Curious Character)*. New York: W. W. Norton & Company, 1985.

Feynman, Richard P. *"What Do You Care What Other People Think?": Further Adventures of a Curious Character*. New York: W.W. Norton & Company, 1988.

Mehra, Jagdish. *The Beat of a Different Drum: The Life and Science of Richard Feynman*. Oxford: Clarendon Press, 2000.

Popova, Maria, "Love after Life: Nobel-Winning Physicist Richard Feynman's Extraordinary Letter to His Departed Wife." Brain Pickings. https://www.brainpickings.org/2017/10/17/richard-feynman-arline-letter/.

Val Fitch

"Cronin and Fitch Detect a Difference between Matter and Antimatter." CERN Timeline, CERN European Council for Nuclear Research, 27 July 1964. Accessed December 9, 2020. https://timeline.web.cern.ch/cronin-and-fitch-detect-difference-between-matter-and-antimatter.

"Hans Courant." Atomic Heritage Foundation, 2020. Accessed 16 December 2020. https://www.atomicheritage.org/profile/hans-courant. "Nobel Laureate and Princeton Physicist Val Fitch Dies at Age 89." Princeton University, 2015. https://www.princeton.edu/news/2015/02/06/nobel-laureate-and-princeton-physicist-val-fitch-dies-age-91.

"The Nobel Prize in Physics, 1980." NobelPrize.org, Nobel Prize Outreach AB, 2021. Accessed December 1, 2021. https://www.nobelprize.org/prizes/physics/1980/fitch/facts/.

"Val Fitch—Interview." NobelPrize.org, Nobel Prize Outreach AB, 2009. Accessed November 30, 2020, https://www.nobelprize.org/prizes/physics/1980/fitch/interview/.

"Val Logsdon Fitch." Princeton University, Department of Physics. Accessed December 1, 2020. https://phy.princeton.edu/department/history/faculty-history/val-fitch.

Fitch, Val. "Val Fitch's Interview." By Cindy Kelly. Manhattan Project Voices National Museum of Nuclear Science & History. March 26, 2008. https://www.manhattanprojectvoices.org/oral-histories/val-fitchs-interview.

———. "Interview of Val Fitch by Finn Aaserud." By Finn Aaserud. Niels Bohr Library & Archives. American Institute of Physics. 1986. https://www.aip.org/history-programs/niels-bohr-library/oral-histories/33299.

Gaulkin, Thomas. "In Their Own Words: Trinity at 75." *Bulletin of the Atomic Scientists* (2020). https://thebulletin.org/2020/07/in-their-own-words-trinity-at-75/.

Overbye, Dennis. "Val Fitch, Who Discovered Universe to Be out of Balance, Is Dead at 91." *The New York Times* (NY), February 10, 2015. https://www.nytimes.com/2015/02/11/us/val-fitch-who-discovered-universe-to-be-out-of-balance-is-dead-at-91.html.

Roy Glauber

McClain, Dylan. "Roy J. Glauber, 93, Dies; Nobel Laureate Explored Behavior of Light." *New York Times*, January 8, 2019. https://www.nytimes.com/2019/01/08/obituaries/roy-j-glauber-dead.html.

"In Memoriam: Roy Glauber." National Museum of Nuclear Science and History, 2020. Accessed 18 November 2020. https://www.atomicheritage.org/article/memoriam-roy-glauber.

"In Memoriam: Roy J. Glauber, 1925-2018." The Optical Society, Updated December 26, 2018. accessed December 8, 2020. https://www.osa.org/en-us/about_osa/newsroom/obituaries/2018/roy_j_glauber/.

"Roy Glauber and His Time in Los Alamos." The Lindau Nobel Laureate Meetings, 2016. Accessed December 9, 2020. https://www.lindau-nobel.org/roy-glauber-and-his-time-in-los-alamos/.

"Roy J. Glauber Member Biography." AIP Publishing, 2020. https://history.aip.org/phn/11511010.html.

"The Nobel Prize in Physics, 2005." NobelPrize.org, Nobel Prize Outreach AB, 2021. Accessed December 1, 2020. https://www.nobelprize.org/prizes/physics/2005/glauber/biographical/.

Dacey, James. "Roy Glauber – from the Bronx to the Nobel." *Physics World*. IOP Publishing, November 18, 2010. https://physicsworld.com/a/roy-glauber-from-the-bronx-to-the-nobel/.

Gingerich, Owen. "Roy Glauber's Interview." Manhattan Project Voices. National Museum of Nuclear Science & History, March 26, 2008. https://www.manhattanprojectvoices.org/oral-histories/roy-glaubers-interview.

McLain, Dylan Loeb. "Roy J. Glauber, 93, Dies; Nobel Laureate Explored Behavior of Light." *The New York Times* (NY), January 8, 2019 2015.

Maria Goepert Mayer

"Maria Goeppert Mayer: Nobel Prize in Physics 1963." The Nobel Foundation, Nobel Prize Outreach AB, 2021. https://www.nobelprize.org/womenwhochangedscience/stories/maria-goeppert-mayer.

Goeppert Mayer, Maria. "On Closed Shells in Nuclei. II." *Physical Review* (American Physical Society) vol. 75, no. 12 (1949): 1969-70. https://doi.org/10.1103/PhysRev.75.1969.

Landau, Elizabeth. "The Last Woman to Win a Physics Nobel." *Scientific American*, September 26, 2017. https://www.scientificamerican.com/article/the-last-woman-to-win-a-physics-nobel1/#.

Goeppert Mayer, Maria. "On Closed Shells in Nuclei. II." *Physical Review* (American Physical Society) vol. 75, no. 12 (1949): 1969-70. https://doi.org/10.1103/PhysRev.75.1969.

Edwin McMillan

"I.I. Rabi (1898 - 1988)."AtomicArchive.com, https://www.atomicarchive.com/resources/biographies/rabi.html.

"Isidor Isaac Rabi: Walking the Path of God." *Physics World*, 1 November 1999. https://physicsworld.com/a/isidor-isaac-rabi-walking-the-path-of-god/.

"Isidor Isaac Rabi" Magnet Academy, National High Magnetic Field Laboratory. https://nationalmaglab.org/education/magnet-academy/history-of-electricity-magnetism/pioneers/isidor-isaac-rabi.

"Man in the News; Humanist Scientist Isidor Isaac Rabi." *The New York Times* (New York), October 28, 1964.

"Part IV: The Manhattan Engineer District in Operation, Los Alamos." AtomicArchive.com. Accessed 22 January 2021. https://www.atomicarchive.com/history/manhattan-project/p4s25.html.

Anderson Jr., O. E., and R. G. Hewlett. *The New World: A History of the United States Atomic Energy Commission*. U.S. Atomic Energy Commission, 1962.

Berger, Marilyn. "Isidor Isaac Rabi, a Pioneer in Atomic Physics, Dies at 89." *The New York Times* (New York), January 12, 1988.

Lofgren, Edward J. "Edwin M. McMillan, a Biographical Sketch." International Symposium: The 50th Anniversary of the Phase Stability Principle, Moscow/Dubna, Russia, Lawrence Berkeley Laboratory, Berkeley, 1994.

Lofgren, Edward J., Philip H. Abelson, and A. Carl Helmholz. "Edwin M. McMillan." *Physics Today* vol. 45, no. 2 (1992): 118.

Isidor Rabi

Rigden, John S. *Rabi: Scientist and Citizen*. New York: Basic Books, Inc., 1987.

"The Nobel Prize in Physics, 1944." NobelPrize.org, Nobel Prize Outreach AB, 2021. https://www.nobelprize.org/prizes/physics/1944/rabi/biographical/.

"Isidor I. Rabi, Biography and Timeline." National Museum of Nuclear Science and History, 2020, accessed 18 November 2020, https://www.atomicheritage.org/profile/isidor-i-rabi.

Britannica, The Editors of Encyclopaedia. "Isidor Isaac Rabi, American Physicist." In Encyclopedia Britannica online. https://www.britannica.com/biography/Isidor-Isaac-Rabi.

Norman Ramsey

"In Memoriam: Norman Ramsey." Brookhaven National Laboratory, 2011. Accessed October 13, 2020. https://www.bnl.gov/newsroom/news.php?a=22732.

"Norman Ramsey." Valley Free Press, 2011. Accessed October 7, 2020. https://www.legacy.com/us/obituaries/vfpnews/name/norman-ramsey-obituary?n=norman-ramsey&pid=154520940.

Chown, Marcus. "Norman Ramsey: Physicist Whose Work Led to the Atomic Clock and the MRI Scanner." *Independent* (London), 2011. https://www.independent.co.uk/news/obituaries/norman-ramsey-physicist-whose-work-led-atomic-clock-and-mri-scanner-6262579.html.

Ramsey, Norman F. "The Atomic Hydrogen Maser." *Meterologia* vol. 1, no. 1 (1965): 7-15.

Ramsey, N. F., "Conferences in Oppenheimer's Office August 12 and 14, 1944 on Air Forces Organization for Our Project."

Ramsey, Norman. "Experiments with Separated Oscillatory Fields and Hydrogen Masers." *Science* vol. 248, no. 4968 (1990): 1612-1619. DOI: 10.1126/science.248.4963.1612.

———. Md-H-1 Form for Norman Ramsey, 1945.

Frederick Reines

"The Nobel Prize in Physics, 1995: Frederick Reines." NobelPrize.org, Nobel Prize Outreach AB, 2021. https://www.nobelprize.org/prizes/physics/1995/reines/biographical/.

Kropp, William, Jonas Schultz, and Henry Sobel. "Frederick Reines, 1918—1998." Washington, D.C.: National Academy of Sciences, 2009.

Kropp, William, M. Moe, L. Price, J. Schultz, H. Sobel, N. J. Teaneck. *Neutrinos and Other Matters: The Selected Works of Frederick Reines*. Hackensack: World Scientific Publishing, 1991.

Joseph Rotblat

Brown, Andrew. *Keeper of the Nuclear Conscience: The Life and Work of Joseph Rotblat*. Oxford: Oxford UP, 2021.

Crosfill, M. L., P. J. Lindop, and Joseph Rotblat. "Variation of Sensitivity to Ionising Radiation with Age." *Nature*, no. 183 (1969): 1729-30. https://doi.org/10.1038/1831729a0.

Hoddeson, Lillian, Paul W. Henriksen, Roger A. Meade, and Catherine Westfall. *Critical Assembly: A Technical History of Los Alamos During the Oppenheimer Years, 1943-1945*. Cambridge: Cambridge University Press, 1993.

Hinde, Robert. A. and J. L. Finney "'Sir Joseph Rotblat', Biographical Memoirs of Fellows of the Royal Society." no. 53 (2007): 309-26.

Lindop, P. and Joseph Rotblat. "Strontium-90 and Infant Mortality." *Nature* 224, no. 5226 (1969): 1257-60. https://doi.org/10.1038/2241257a0.

Rotblat, Joseph. "Joseph Rotblat's Interview." By Martin J. Sherwin. Voices of the Manhattan Project. National Museum of Nuclear Science and History. 1989. www.manhattanprojectvoices.org/oral-histories/joseph-rotblats-interview.

———. "Emission of Neutrons Accompanying the Fission of Uranium Nuclei." *Nature*, no. 20 (1939).

———. "Induced Radioactivity of Nickel and Cobalt." *Nature*, no. 28 (1935).

Emilio Segrè

"Emilio Segrè – Facts." NobelPrize.org, Nobel Prize Outreach AB, 2021. Accessed December 2, 2020. https://www.nobelprize.org/prizes/physics/1959/segre/facts/.

"The Bevatron: 40 Years of Science." Berkeley Lab, 2021. Accessed December 2, 2020. https://berkeleylab.exposure.co/the-bevatron.

Anderson, O. E. Jr. and Hewlett, R. G. *The New World: A History of the United States Atomic Energy Commission*. The U.S. Atomic Energy Commission, 1962.

Chadwick, James. "The Existence of a Neutron." Proceedings of the Royal Society of London. Series A, Containing Papers of a Mathematical and Physical Character 136, no. 830 (1932): 692-708.

Flint, Peter B. "Dr. Emilio G. Segrè Is Dead at 84; Shared Nobel Studies of Atoms." *The New York Times* (New York), 1989. https://www.nytimes.com/1989/04/24/obituaries/dr-emilio-g-segre-is-dead-at-84-shared-nobel-for-studies-of-atom.html.

Hawkins, David and Edith Truslow. *Manhattan District History Project Y: The Los Alamos Project*. Los Alamos, New Mexico: Los Alamos Scientific Laboratory, 1946.

Hoddeson, Lillian, Paul W. Henriksen, Roger A. Meade, and Catherine Westfall. *Critical Assembly: A Technical History of Los Alamos During the Oppenheimer Years, 1943-1945*. Cambridge: Cambridge University Press, 1993.

Segrè, Emilio. *A Mind Always in Motion: The Autobiography of Emilio Segrè*. Oakland: University of California Press, 1993.

"What were they working on?"

Carr, Alan. "Why Wasn't Little Boy Tested?," *LANL Inside*, August 2020.

"Establishing Los Alamos." The Manhattan Project - an Interactive History. U.S. Department of Energy. https://www.osti.gov/opennet/manhattan-project-history/Events/1942-1945/establishing_los_alamos.htm.

"J. Robert Oppenheimer." National Museum of Nuclear Science and History, Updated July 7, 2014. https://www.atomicheritage.org/profile/j-robert-oppenheimer.

"Little Boy and Fat Man." National Museum of Nuclear Science and History, Updated July 23, 2014. https://www.atomicheritage.org/history/little-boy-and-fat-man.

"Oppenheimer Security Hearing." National Museum of Nuclear Science and History, updated July 7, 2014. https://www.atomicheritage.org/history/oppenheimer-security-hearing.

"The Manhattan Project." National Museum of Nuclear Science and History, 2020. https://www.atomicheritage.org/history/manhattan-project.

"World War II 1939-1945." Britannica, updated Aug 27, 2021. https://www.britannica.com/event/World-War-II.

Krebs, Albin. "Gen. Groves of Manhattan Project Dies." *The New York Times* (New York), July 15, 1970. https://www.nytimes.com/1970/07/15/archives/gen-groves-of-manhattan-project-dies-gen-groves-of-manhattan.html.

National Security Research Center Staff. "Witnessing Trinity: The Test's Success Was Met with Surprise, Relief, Fear," *LANL Inside*, June 2020.

Whitacre, Madeline and Alan Carr. "Seventy-Five Years Ago, Los Alamos Scientists Detonated the World's First Nuclear Explosion," *LANL Inside*, June 2020.

"Why didn't Oppie ever win a Nobel prize?"

"J. Robert Oppenheimer: Life, Work, and Legacy." Institute for Advanced Study, Updated 2021, https://www.ias.edu/oppenheimer-legacy.

"J. Robert Oppenheimer, Atom Bomb Pioneer, Dies; Physicist Cancer Victim Censured and Later Honored by the A.E.C. Dr. J. Robert Oppenheimer, 'Father of the Atomic Bomb,' Dies in Princeton Famed Physicist Was Long Ailing Career Had Been Marked with Controversy since Hearings of 1953." *The New York Times* (New York), January 8, 1967. https://www.nytimes.com/1967/02/19/archives/jrobert-oppenheimer-atom-bomb-pioneer-dies-physicist-cancer-victim.html.

Bird, Kai and Martin J. Sherwin. *American Prometheus: The Triumph and Tragedy of J. Robert Oppenheimer*. New York: Alfred A. Knopf, 2005.

Cassidy, David C. *J. Robert Oppenheimer and the American Century*. Baltimore: John Hopkins University Press, 2009.

Cherniss, Harold. "Harold Cherniss's Interview." By Martin J. Sherwin. Manhattan Project Voices National Museum of Nuclear Science & History. May 23, 1979. https://www.manhattanprojectvoices.org/oral-histories/harold-chernisss-interview-part-1.

Christys, Robert. "Robert Christy's Interview." By Martin J. Sherwin. Manhattan Project Voices National Museum of Nuclear Science & History. March 30, 1983. https://www.manhattanprojectvoices.org/oral-histories/robert-christys-interview.

Glauber, Roy. "Roy Glauber's Interview" by Martin J. Sherwin. Manhattan Project Voices National Museum of Nuclear Science & History. March 26, 2008. https://www.manhattanprojectvoices.org/oral-histories/roy-glaubers-interview.

Groves, Leslie. *Now It Can Be Told: The Story of the Manhattan Project*. New York: Harper & Brothers, 1962.

Monk, Ray. *Robert Oppenheimer: A Life Inside the Center*. New York: Anchor Books, 2014.

Morgan, Thomas B. "With Oppenheimer on an Autumn Day." *Look* vol. 30, no. 26 (December 19 1966).

Templeton, Patty, "Plutonium and Poetry: Where Trinity and Oppenheimer's Reading Habits Met," *LANL Inside*, 2021.

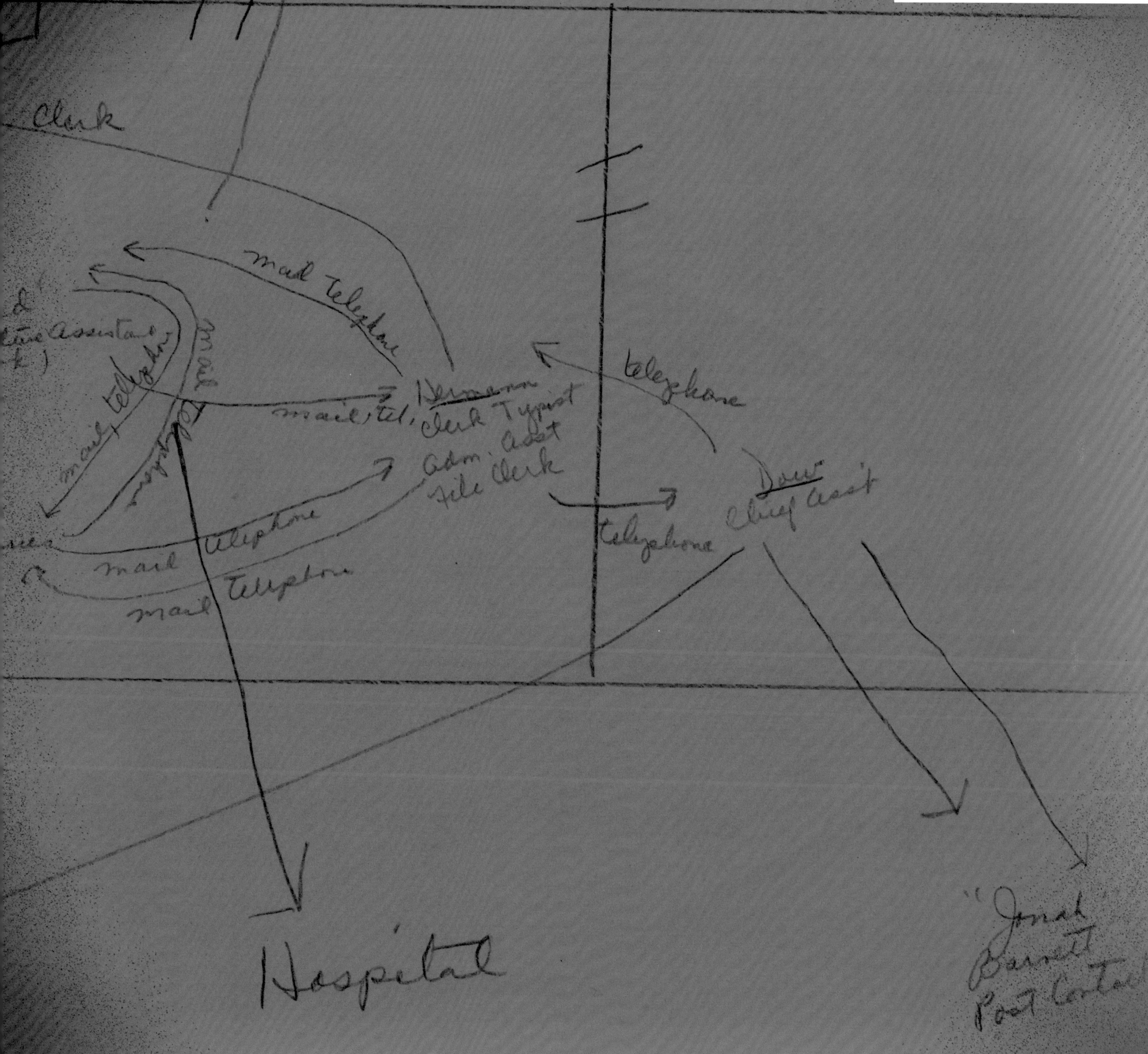
Clerk
Mail Telephone
Mail Telephone
mail, telephone
mail, tel,
Hermann
Clerk Typist
Adm. Asst
File Clerk
telephone
Telephone
Dow
Chief Ass't
mail Telephone
mail Telephone
Hospital

Contributors

Amy Belotti is a digitizer-archivist at the National Security Research Center. She has a bachelor's degree in film and a master's degree in library and information science. She supports the digitization of multiple formats at the lab, specializing in the care and digitization of audiovisual materials. Belotti is from New York state. She misses good bagels, but breakfast burritos with Christmas chile will suffice.

Alan B. Carr is the senior historian and a program manager at the National Security Research Center at Los Alamos National Laboratory. He began his career at the lab in 2003 after completing a master's degree in history at Texas Tech University. Carr appraises and is the formal custodian of millions of films, photo negatives, and documents that form LANL's vast collection of records. He has lectured for professional organizations, has been a guest on national and international radio and television programs, and has produced publications on top-secret weapons from World War II to the Cold War.

Hadley Hershey is the National Security Research Center's digital collections lead digitizer-archivist. They are responsible for leading multiple digitization projects, interacting with customers, developing cost estimates, and training staff to digitize and perform quality control reviews for various types of media. Hershey came to the lab with a decade of archives experience, including managing the digital collections program at a law library. Hershey has a bachelor's degree in history and a master's degree in library and information science with a concentration in archives and records management. They are also a certified archivist and digital archives specialist. A Pennsylvania native, Hershey grew up on a farm 40 miles from Hershey, Pennsylvania.

Jackie Kilby is a former digitizer-archivist at the National Security Research Center, where she digitized some of the lab's most important historical material. Her passion has always been history, specifically pre-colonial and American history. Her education consists of a bachelor's degree in history and a master's degree in library and information science, focusing on archiving and cultural heritage institutions. A born-and-raised Baltimorean, Kilby loves Old Bay seasoning and the Orioles baseball team, but does not miss East Coast traffic.

Megan Jochem is an archives coordinator responsible for the physical collections and their locations held by the National Security Research Center. Her education includes a bachelor's degree in history, a master's degree in museum studies, and a master's degree in library and information science. Jochem's professional experience includes work at the National Archives and Records Administration in Washington, D.C., where she digitized records from the War of 1812, as well as at the National Archives (Washington National Records Center). Jochem grew up in Sacramento, California, where she would help her dad harvest walnuts on their farm and swim year-round.

Nicholas Lewis is a historian who uses the National Security Research Center's extraordinary resources to explore how the scientific and technical achievements of Los Alamos have helped shape world history since the Second World War. He has long held a fascination with how human societies and technological systems mutually construct one another. With undergraduate degrees in history and anthropology, a master's degree in world history, and a Ph.D. in the history of science and technology, Lewis wrote his dissertation on the unique development of supercomputing at Los Alamos during the Cold War. Lewis is a native of Bountiful, Utah, and is a collector of antique adding machines and typewriters.

Mott Linn was the National Security Research Center's chief librarian until his recent retirement. While the chief librarian, he oversaw the acquisition, organization, and accessibility of the collections. Linn was a career-long archivist with a distinguished service award from the Academy of Certified Archivists. He has a doctorate in library management from Simmons University, and master's degrees in history, library science, and nonprofit management. His bachelor's degree is in history. His hometown is Harleysville, Pennsylvania.

Ellen McGehee is a National Security Research Center historian with a Ph.D. in history from the University of New Mexico. She moved to Los Alamos at the age of 13 and, when asked, would say that northern New Mexico is her real home. She started her career at LANL as a newly-minted archaeologist, having received her bachelor's degree in anthropology. Completing her graduate studies in history while working at the lab, she has researched the prehistoric and historic past, specializing in the Homestead, Manhattan Project, and Cold War periods. McGehee formally retired from the lab in 2018 and returned in 2020 as a part-time historian with the National Security Research Center. Her passion for historic preservation contributed to the establishment of the Manhattan Project National Historical Park unit at Los Alamos, which was the highlight of her career. She considers herself a true "Labbie," coming from a family with three generations of LANL workers.

Laura McGuiness is an archivist at the National Security Research Center. She started at the lab in July 2020. McGuiness has a bachelor's degree in English literature from the University of California, Berkeley, and a master's degree in library and information science with an emphasis in metadata and cataloging from San Jose State University. McGuiness's work focuses on populating metadata for digital and physical collections. She is a California native and an animal lover, who volunteered weekly for seven years at a wildlife rehabilitation center, bottle-feeding baby raccoons, giving injections to owls, and gleefully throwing lettuce to ducklings.

Roger Meade is a retired laboratory archivist and historian now working as a historian for the lab's Chemistry Division's Radiochemistry Assessment Team and the National Security Research Center. Among his degrees, Meade holds a master's degree in history from Wright State University in Ohio, and a Ph.D. in history from Arizona State University. Hired at the lab in 1984, Meade expanded the nascent archives into an unparalleled resource for both scientific and historical research. He later became a co-author of Critical Assembly, a history of the laboratory's scientific and technical accomplishments during World War II, working with Edward Teller, Hans Bethe, and Norris Bradbury. More recently, Meade assisted in the film production of *The Half-Life of Genius, the life and times of physicist Raemer Schreiber*. After his initial retirement from the lab, Meade worked for Arizona State University before returning to Los Alamos.

Renae Mitchell is a communications specialist with a Ph.D. in comparative literature from Penn State University. She joined the lab in 2020, and is a writer and editor for internal and external laboratory publications and for the National Security Research Center. Writing and teaching proper and creative usage of the English language has driven much of Mitchell's career, from teaching English in Japan to teaching at the university level for over 10 years. Mitchell has published scholarly articles in her field, and her first book was published in 2021. She is a Louisiana native, and tries to live by her home state's motto: "laissez les bon temps rouler!"

John Moore is a media-archivist/historian with the National Security Research Center. Moore has a bachelor's degree in history from Liberty University in Lynchburg, Virginia and a master's degree from Arizona State University in Tempe. He first joined the Los Alamos lab as a student in Records Management. Today, Moore specializes in the photo negative, motion picture, and media collections, which date from the Manhattan Project to the early 2000s.

Richard Moore is the historian for the Atomic Weapons Establishment, which designs, manufactures, and maintains the warheads for the United Kingdom's nuclear deterrent as well as works to support nuclear threat reduction. He was born in Cleethorpes, a seaside resort on the east coast, and educated at both of Britain's great ancient universities: the University of Cambridge and the University of Hull. Moore has a master of arts, master of philosophy, and doctorate degrees and has published two books and numerous journal articles on nuclear history. He lives with his wife and two children in Gloucestershire, where he is fighting a losing battle with the bindweed in his garden.

Mike Nudelman leads the Los Alamos lab's Visual Design and Web teams. He fondly remembers geeking out over stories about Manhattan Project scientists, such as Richard Feyman, as an undergrad at Cornell University — though he never thought his fine arts degrees would (or could!) lead to a dream job in Los Alamos. Nudelman grew up on Long Island, New York, but will take northern New Mexico's mountains over the beach any day.

Octavio Ramos, Jr. is a communication specialist and has more than 32 years of experience. In that time, Ramos has helped various subject-matter experts write books on topics like describing cyclodextrins (complexing agents used in pharmaceuticals), analyzing expert judgement (interpreting and validating data), and chronicling New Mexico's Cerro Grande Fire in 2000. Ramos has also published numerous short stories and five novels, as well as nonfiction books on various styles of martial arts. His education includes a bachelor's degree in professional writing, as well as graduating from the New Mexico Mounted Patrol police academy. He was born in the UFO capital of the world, Roswell, New Mexico, but has never had a close encounter.

Gabriella Smith is a LANL visual designer and animator, who engages in every dimension of design. She uses her creative passions to personify and illustrate exciting new scientific ideas. As a New Mexico Highlands University media arts student, she completed several undergraduate internships with organizations, such as LANL's Bradbury Science Museum and her university. Smith earned her bachelor's degree in the spring of 2021. In her free time, she enjoys making waffles and bringing ordinary objects to life with glue-on googly eyes.

Brye Steeves is a communications specialist with the enviable job of sharing some of history's most-fascinating stories through the National Security Research Center's products, like *The Vault* magazine, *Relics* podcast, mini documentary videos, and books just like this one. Her greatest love will always be writing, which led her to a career as an award-winning newspaper reporter and has provided the foundation for all forms of storytelling since, including authoring a children's book and a dissertation on renewable energy in Portuguese. Her education includes a bachelor's degree in journalism and a master's degree in international relations. A Boise, Idaho native, Steeves did not grow up on a potato farm, but does love French fries.

Madeline Whitacre is an archivist and historian with the National Security Research Center. She joined the laboratory in 2017 as a student. Whitacre's projects include researching and writing about the history of the laboratory, and sharing those stories through presentations and publications. Whitacre grew up in Los Alamos and attended the University of New Mexico, where she received a bachelor's degree in history. She is now pursuing a master's degree in museum profession at Syracuse University in New York.

Paul Ziomek is the National Security Research Center's art director. Originally from Chicago, Ziomek earned an associate's degree in advertising art and a bachelor's degree in illustration while living in Albuquerque. He has developed publications for the City of Roswell, OppenheimerFunds, and many organizations within LANL. Ziomek continues to self-publish graphic novel projects and is a founding member of the nonprofit New Mexico comic book creators' group 7000BC.

Index

W

Z

6.0 SEC.

N

⊢⊣ 100 METERS